D1515009

Ultraviolet Reflections
Life under a Thinning Ozone Layer

Annika Nilsson

REMOVED FROM THE
ALVERNO COLLEGE LIBRARY

551.514
N 712

JOHN WILEY & SONS

Chichester · New York · Brisbane · Toronto · Singapore

Alverno College
Library Media Center
Milwaukee, Wisconsin

Copyright © 1996 by John Wiley & Sons Ltd,
Baffins Lane, Chichester,
West Sussex PO19 1UD, England

National (01243) 779777
International (+44) (1243) 779777
e-mail (for orders and customer service enquiries): cs-books@wiley.co.uk
Visit our Home Page on http://www.wiley.co.uk
or http://www.wiley.com

All Rights Reserved. No part of this book may be reproduced, stored in a retrieval system, or transmitted, in any form or by any means, electronic, mechanical, photocopying, recording, scanning or otherwise, except under the terms of the Copyright, Designs and Patents Act 1988 or under the terms of a licence issued by the Copyright Licensing Agency, 90 Tottenham Court Road, London, UK W1P 9HE, without the permission in writing of the publisher.

Other Wiley Editorial Offices

John Wiley & Sons, Inc., 605 Third Avenue,
New York, NY 10158-0012, USA

Jacaranda Wiley Ltd, 33 Park Road, Milton,
Queensland 4064, Australia

John Wiley & Sons (Canada) Ltd, 22 Worcester Road,
Rexdale, Ontario M9W 1L1, Canada

John Wiley & Sons (Asia) Pte Ltd, 2 Clementi Loop #02-01,
Jin Xing Distripark, Singapore 129809

Library of Congress Cataloging-in-Publication Data
Nilsson, Annika.
    Ultraviolet reflections: life under a thinning ozone layer/
Annika Nilsson.
        p.   cm.
    Includes bibliographical references and index.
    ISBN 0-471-95843-3 (pbk.)
    1. Ozone layer depletion.   2. Ultraviolet radiation.   I. Title.
QC879.7.N55   1996
363.73'84—dc20   96-11463
                                                          CIP

British Library Cataloguing in Publication Data

A catalogue record for this book is available from the British Library

ISBN 0-471-95843-3

Typeset in 11/13pt Baskerville from the author's discs by Vision Typesetting, Manchester
Printed and bound in Great Britain by Biddles Ltd, Guildford and King's Lynn
This book is printed on acid-free paper responsibly manufactured from sustainable forestation, for which at least two trees are planted for each one used for paper production.

Ultraviolet Reflections

Contents

Preface and acknowledgements

MY JOURNEY INTO THE WORLD OF PHOTOBIOLOGY started in the fall of 1992 with an invitation to attend a conference in Sardinia. The Scientific Committee on Problems of the Environment (SCOPE) was in the process of formulating a state-of-the-art report and research plan on the effects of increased ultraviolet radiation on global ecological systems under the chairmanship of Professor Edward de Fabo. The meeting brought me in contact with some of the key researchers in the field. It also inspired me to dig deeper into the effects of ozone depletion. The atmospheric science of ozone depletion had been a field of journalistic interest for me for many years, including the politics of CFC phase-out. It surprised me that so little was known about the biological implications of this global environmental change.

About six months after the meeting, I received a letter from marine biologist Patrick Neale inviting me to participate in a research cruise to Antarctica. It was an opportunity I could not pass by. His grant from the US National Science Foundation came to pay for my stay on board the research vessel and ice-breaker *Nathaniel B. Palmer*. A grant from the Swedish Natural Science Research Council paid for my travels to our point of departure, Punta Arenas, Chile.

The trip to the ozone hole in the fall of 1993 became the starting point for realizing the idea of a book. The scientists on board, the staff from the Antarctic Support Associates and the crew of the *Nathaniel B. Palmer* all provided information and inspiration for my writing. I also owe my gratitude to the sea, the ice, the light, the wind and the waves that gave the surroundings for the chapter on plankton life under the ozone hole.

Earlier that fall, I had given myself the opportunity to travel north in Sweden to visit the Abisko Scientific Research Station. It is an area in which I have both skied and hiked and have always enjoyed the spectacular landscape. Two of the researchers at the Station, my guides during the visit, Carola Gehrke and Ulf Johanson, gave me a new focus, on the low alpine brush and the moss that covers the ground. Together with Professor Lars Olof Björn at their home university in Lund, they helped provide information and comments about the effects of UV-B on plant life.

Already on Sardinia, I had made connections with Edward de Fabo and Frances Noonan, and in the fall of 1994 I had an opportunity to visit them at their laboratory in Washington, DC. Aside from providing me with much of the background material on immunological effects of UV-B, they inspired me to think about the evolutionary implications of living with the sun as a constant companion.

For freelance writers, the financial situation has a tendency to determine one's rate of progress in writing. In the fall of 1994, I was lucky to receive a stipend from the Swedish Cancer Foundation, which allowed me to put time into gathering information about cancer and ultraviolet radiation. Their interest was in the connection between solariums and malignant melanoma. There is a certain irony in that some people freewillingly expose themselves to extra ultraviolet radiation at the same time as we worry about ozone depletion and skin cancer.

Gathering information about cancer brought me to Professor Jan van der Leun and some of his colleagues in Utrecht. It was an inspiring visit, and throughout my writing Jan van der Leun has been of great help in explaining the mechanisms of cancer and in judging some of the evidence. Moreover, he gave me some important insights into the process of risk assessment.

My thinking about the process of science when it intersects with politics was further stimulated by science philosopher Jan Nolin in Gothenburg.

For many of the chapters, I have relied on written material, including the reports from the Environmental Effects Panel within the United Nations Environment Programme. My text is not footnoted, but I want to acknowledge that without the research, the articles, the reviews and the discussions within the scientific community, this book could not have been written.

Several people have helped me review parts of the text. Aside from the scientists already mentioned, they are Professor Henning Rodhe at Stockholm University, Weine Josefsson at the Swedish Meterological and Hydrological Institute, and Professor Barbara E. K. Klein at the University of Wisconsin. I am grateful for their useful comments.

Gathering the material is one part of writing a book. The other part of this project has been getting everything down on paper—a slow struggle between me and my computer. Without the support and encouragement of my friend and partner Cindy de Wit, the words would still be floating around in my head. She has been my patient reader and language editor, and she has kept me going even when the going has been rough.

I write this preface a few days before the winter solstice of 1995. It is close to the summer solstice in Antarctica. It reminds me that all journeys are spiral. Rather than merely passing from one point to another, from Sardinia via Abisko and Antarctica and back to my home in Huddinge, Sweden, the journey of writing this book is about looking at life from constantly new perspectives. By spinning further, the picture will shift slightly, which is the process of discovery. I hope this book conveys the importance of looking at science as a process and that you as a reader feel invited to participate in the continued journey.

Chapter 1

From Abisko to Antarctica

L ATE AUGUST IN SWEDEN. IN THE MOUNTAINS NORTH of the Arctic circle, the wind-blown birch forest is still green, but the first signs of fall have just appeared—a few yellow leaves, a nip in the air and new snow on the mountain tops. At Abisko Scientific Research Station, two students are busy at work. There are only a few weeks left of the field season before snow will cover the low brush and moss that form the floor of the thin forest. They measure and weigh shoots and leaves of the plants to search for an effect of ultraviolet radiation on the growth. The students suspect that this fragile plant life might be especially sensitive to any changes in the light environment caused by a depletion of the ozone layer.

Two months later at the opposite pole. It is late October in the Weddell–Scotia Sea in Antarctica. Spring is on its way and the light has come back after a long dark winter. The sea ice breaks up as the solar energy is absorbed and then helped by wind and waves. The broken-up ice creates crevices of life and the colors shift from white to reddish brown as ice algae multiply. Soon the sea is free. Fresh cold water from the melting ice flows northward. Light and nutrients form the base of a spring bloom at sea. But the light is both a blessing and a curse. Since the ozone hole was discovered over Antarctica in the mid-1980s, many people have been concerned that damaging ultraviolet radiation will reach the sensitive sun-catching cells in the sea. Will the spring bloom suffer? Will there be less food for the Antarctic wildlife? No one knows what long-term effects the ozone hole will have on marine life, but the signs are gathering and they do not look good.

Abisko and the Weddell–Scotia Sea are both far away from any industries or direct pollution. Yet, the pictures from north and south remind us that the effects of human activities do not always care about physical distance. Decades of emissions of man-made stable chemicals have created a truly global threat—the depletion of the ozone layer that normally protects life from damaging solar rays.

A HISTORICAL BACKGROUND

The ozone layer first came into the environmental debate in the 1970s, when the United States and France were developing supersonic airplanes. Atmospheric chemist and 1995 Nobel Laureate Paul Crutzen had recently pointed out the role of nitrogen oxides in the natural destruction of ozone, and the fear was that exhaust from the airplanes would spew nitrogen oxides directly into the upper parts of the atmosphere. The plans for a large fleet of supersonic airplanes were eventually put on the shelf, but in the meantime a new threat had become apparent—man-made chlorine-containing chemicals, the chlorofluorocarbons or CFCs for short.

CFCs have a long industrial history. They were originally developed in the 1930s as a modern, safe coolant for refrigerators. Since then they have found applications in the production of insulating foam, for cleaning circuit boards and as propellants in spray cans, to mention only a few examples. In the 1970s, their production was skyrocketing because they were such versatile and useful chemicals. No one worried about releasing them into the environment. They were stable, non-toxic and thought to be very safe.

The turn-around came in 1974 when the two chemists Sherwood Rowland and Mario Molina proposed that chlorine from these chemicals could break loose and act as a potent destroyer of ozone—a theory that later led to their receiving the Nobel Prize in chemistry together with Paul Crutzen. Earlier studies using CFCs as tracers of air movement showed that they, unlike many less stable chemicals, could indeed reach the stratosphere.

Rowland's and Molina's claim was more than a scientific theory. It was an environmental warning that threatened a powerful chemical industry and it immediately kicked off a very heated debate among environmentalists and industrialists alike. Consumers soon got involved by boycotting spray cans as an easily identified offending product. It was a veritable "Ozone War", as has been described by science writer Lydia Dotto. Environmentalists and the chemical industry both claimed that their version of the truth and their risk assessment was the most accurate. Science and scientists played major roles in the debate. In some countries, the discussion eventually led to a ban on CFCs as propellants in spray cans.

CFCs know no national boundaries and in spite of scientific disputes, the issue was quickly carried onto the scene of international politics with the United Nations Environment Programme and the World Meteorological Organization as major actors. The discussions led to the Convention for the Protection of the Ozone Layer, which was adopted in Vienna in March 1985. However, as everyone thought there was ample time to deal with the problem, the Vienna Convention did not make any specific demands on the signatories.

Two months later came the shock. Measurements from a British research station at Halley Bay in Antarctica showed that there had been severe depletion of ozone during the Antarctic spring—the ozone hole (Figure 1.1). The destruction of ozone was much more severe than had been predicted by anyone and the scientist in charge, Joseph Farman, was not even sure whether to trust his instrument. In another series of measurements of the ozone layer, by an American satellite, the extremely low values had been sorted out as anomalies. It was not until Farman asked them to go back to the primary

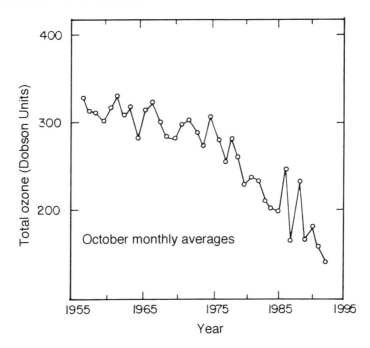

Figure 1.1   The springtime measurements of total ozone for Halley Bay, Antarctica.
These measurements were the alarm clock that alerted the world to the ozone hole.
Source: *Scientific Assessment of Ozone Depletion: 1994*

record that the Halley Bay measurements could be confirmed. The extremely
rapid destruction of ozone was a nasty surprise for everyone and it took several
years before the chemical processes in the Antarctic atmosphere could be
adequately explained.

In spite of the remaining scientific uncertainties, the international political
process started by the Vienna Convention was put in high gear. In 1987, the
negotiations resulted in a commitment to reduce the production and emission
of the damaging chemicals. The agreement, called the Montreal Protocol, has
since been revised several times. The current Protocol requires all industrialized
nations to completely phase out CFCs by 1996. The Protocol also regulates
other damaging substances that contain chlorine and bromine. Developing
countries have a 10-year grace period for getting rid of their ozone-depleting
substances, but the international goal is nevertheless very clear—to save the
ozone layer even if it requires major and sometimes costly changes in many
important industries. The efforts have borne fruit. Many companies have
successfully converted to ozone-friendly technologies (Figure 1.2). Recently

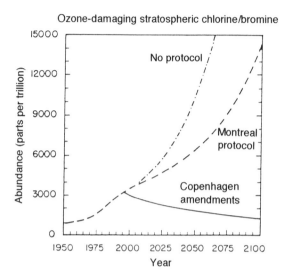

Figure 1.2   Without the political commitments to phase-out CFCs in the Montreal Protocol and its amendments, the chlorine concentration in the atmosphere would skyrocket. Source: *Scientific Assessment of Ozone Depletion: 1994*

the increase of CFCs in the atmosphere has started to level off, which will eventually lead to a recovery of the ozone layer.

BACKLASH IN THE MIDST OF ACTION

By now, the threat to the ozone layer is common knowledge in most of the industrialized world, even if most people only have a very general picture of the problem. Consumers ask for CFC-free products such as refrigerators and the producers are starting to meet the demand. Many have read about increased risks of skin cancer, and indexes warning about ultraviolet radiation are becoming part of the usual weather forecast. Sunblockers are selling well during summer months. In the developing world, the awareness of the need to phase out ozone-destructive technologies is rising as well.

At the same time, louder and louder voices are saying that there is no threat to the ozone layer—that the whole problem of CFCs and increases in ultraviolet radiation is a myth created by scientists to get attention and money for their research. Books such as *The Holes in the Ozone Scare* by Rogelio Maduro and Ralf Schauerhammer use arguments from the scientific literature to get their point across.

Their influence is not negligible. Many people have started to question the reality of the ozone hole. "But I have heard it is not a real problem," several well-educated people told me after I had returned from Antarctica in November 1993—when the worst depletion on record had just been recorded. In Arizona, the state government has been discussing whether to allow CFCs again, so as not to threaten the continued use of air conditioners. This would be contrary to both international agreements and federal US policy. During a Congressional hearing in the fall of 1995, leading Republicans questioned the scientific theory behind ozone destruction by CFCs, calling the process of peer review and scientific consensus "mumbo-jumbo." Internationally, smuggling of CFCs is undercutting efforts to phase out these substances.

JUDGING THE EVIDENCE

How should anyone not immediately involved in the field of ozone research judge the arguments put forth in the backlash against the evidence that forms the basis for the international agreement? With great difficulty, I would say. It takes effort to read the primary scientific literature. The alternative is to rely on the consensus reports that the United Nations Environment Programme (UNEP) produce as a way to follow up the Vienna Convention. The committee responsible for the scientific assessments has in its most recent report tried to answer some of the common claims by the critics. The language is aimed at a lay audience and the text is worthwhile reading for anyone trying to understand the different points of view in the controversy. Their conclusion is very clear. Ozone depletion is a reality and the main culprits are man-made chlorine chemicals, such as CFCs.

But what does ozone depletion mean? Maybe the threat from this change in the atmosphere has been exaggerated? The answers to such questions are not as clear-cut, mostly because the research on the effects of increases in ultraviolet radiation is lagging far behind the atmospheric sciences. There are some answers, however, and the Environmental Effects Panel within UNEP regularly publishes a report based on a consensus among all the leading scientists in the field. The most recent report reaffirms that there are some real health effects, such as an increased risk of skin cancer, as well as many potential problems for natural ecosystems both on land and in the sea.

The ozone debate and the political decisions about phasing out ozone-depleting substances have been built on a trust on experts. However, in the current discussion climate scientific consensus does not seem to be enough. The science itself is being questioned. The problems get accentuated when sacrifices from industry start costing money and when myths about ozone depletion take on a life of their own. In this situation there are two choices. One is to intuitively trust the consensus reports or the critics. The other is to gain a better understanding of the questions at hand by learning more about what is happening in the sky and how it can affect health and subsistence as well as the natural environment.

I believe that we also need to understand the limitations of science and the difficulties in answering all the questions. What scientists can and cannot predict. Some of the "truths" presented in the mass media are far from clear-cut when one takes a closer look at the evidence. There have been warnings that create unnecessary fear and others that have been misguided because of a lack of understanding of the biology involved. Other "truths" have been proven wrong with new observations and as new insights have emerged.

INVITATION TO A JOURNEY

This book is about trying to understand the damage that can be caused by ozone depletion. It is about the effects of ultraviolet radiation on life. Most popular science publications focus on facts. This is also a book of questions and an invitation to follow my journey in trying to find some answers. It is a book about science in progress, trying to catch up with, record and predict the damage caused by ozone depletion.

What becomes clear is that some structures and processes that are important for life are very sensitive to ultraviolet radiation and could be hit hard by ozone depletion. Others are more robust as life has learned to adapt. In the end, it often becomes a question of what we mean by risk. The study of ultraviolet-related research also offers some insights into the process of science—the pitfalls, the controversies and the difficulty in seeing the whole picture.

The questions I raise should not only concern the experts and professionals who have ozone issues as part of their job. Some of them might find my treatment of the subject too shallow. Instead the book is aimed at all those who are concerned without having the time or capacity to follow the scientific literature. You could be a science teacher trying to convey basic ideas about the environment, a student who wants to know things that you cannot find in the textbooks, an environmentalist or a policy maker needing more than the statements of scientific consensus. Maybe you are just curious to learn more about one of the major environmental issues of today.

I should issue a warning. You will not find simple answers to the questions we often ask. Is it dangerous or not? Truth is seldom that simple and more often than not it is hidden. My aim is instead to convey some of the complexity of living systems and the difficulty in predicting their behavior and in assessing risk. Why? Because I believe that an appreciation of the realities of science is important for an informed environmental discussion in society—a discussion that does not only involve the experts. It is also because understanding biology is a joyous part of understanding life.

Chapter 2

Let there be light

In the beginning of creation, when God made heaven and Earth, the Earth was without form and void, with darkness over the face of the abyss, and a mighty wind swept over the surface of the waters. God said, "Let there be light." And there was light; and God saw that the light was good.

THIS JUDEO-CHRISTIAN LEGEND IS MIRRORED IN MANY other cultures. Sometimes the sun is created from the yolk of an egg. Often she has been seen as the Great Mother Goddess. Almost always the sun is symbolic of life. This is no coincidence. Sunlight is a prerequisite for life on Earth.

But the sun has a darker side as well. The energy from the solar rays can disrupt the hereditary material in living cells, the very molecules of life. This destructive potential has always been there and life has had to adapt. Plants, which harvest light to thrive, have to shield themselves with pigments and thick leaves against getting too much. Plankton dive to escape when the radiation becomes too intensive. People living close to the equator take siestas when the sun is highest in the sky.

For millions of years the ozone layer has served as an effective shield against the most damaging radiation emitted by the sun. However, the changing composition of our atmosphere has put the darker side of the sun into new focus. This chapter looks at the physical nature of this threat. It is a background to better understand how the solar rays interact with the atmosphere to determine the light environment on Earth.

ENERGY AS WAVES AND PHOTONS

Sunlight is radiant energy. This energy is packaged as photons that can be captured and altered into other kinds of energy, such as chemical bonds and heat. Some of this energy creates the warmth we feel on our skin on a sunny day. It makes it possible for our eyes to record the world around us as the sun lights up the sky. And it is this energy that makes plants grow as they convert the light into sugars and store it in leaves, stems and roots.

Sunlight can also be described in another way. Its colorless radiance reveals this second nature when the light is split into a spectrum of colors by a prism or by raindrops in a rainbow. The colors, ranging from violet to red, are light of various wavelengths. Each of these wavelengths represents photons with different levels of energy. The shorter the wavelength, the higher the energy level of each photon. Photon for photon, blue light is richer in energy than green, which is more energetic than red, etc.

Some parts of the spectrum are not visible (Figure 2.1). Further away at the red end of the spectrum is infrared—heat. At the opposite end of the visible spectrum, just outside the violet end, is ultraviolet (UV) radiation. This book will focus on UV radiation and its interaction with living things, in particular how

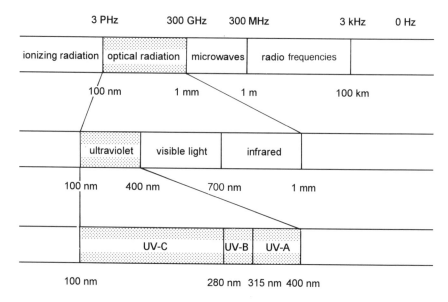

Figure 2.1   The electromagnetic spectrum. Source: *UV Radiation from Sunlight*

life has evolved or not evolved to deal with the highly energetic photons in these wavelengths.

THE FATE OF PHOTONS WITH A NOTE ON TERMINOLOGY

About 9 percent of solar radiation is in the UV wavelengths. That is not what reaches Earth, however. On their journey, the photons will meet many different molecules that can absorb or scatter the energy. Some of the photons are scattered by the nitrogen and oxygen molecules of the air and reflected back into space. Others are absorbed, especially by oxygen and by ozone. This way the atmosphere effectively filters out all wavelengths that are shorter than 280 nanometers (abbreviated $nm = 10^{-9}$ meters). Photons with wavelengths between 280 and 315 nm are effectively absorbed by ozone and only rarely reach all the way through the atmosphere. As the wavelengths increase, more photons will pass unhindered. Ultraviolet light of wavelengths longer than 320 nm are hardly affected at all by ozone and follow the same path as visible light (Figure 2.2).

Corresponding roughly to this behavior in the atmosphere, the UV radiation is commonly divided into three parts depending on the wavelength: UV-A (400–315 nm), UV-B (315–280 nm) and UV-C (280–100 nm). The

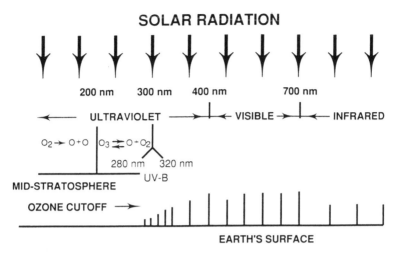

Figure 2.2   Some of the solar radiation is filtered by the atmosphere. The ozone layer in the mid-stratosphere effectively cuts off any radiation of wavelengths less than 280 nanometers. Source: *Effects of Increased Ultraviolet Radiation on Biological Systems*

division is somewhat arbitrary as the cut-off points are not as sharp in reality. There are also other ways of categorizing the different parts of the UV spectrum, such as extreme UV (10–100 nm), far UV (100–180 nm), middle UV (180–300 nm) and near UV (300–380 nm).

A CLOSER LOOK AT OZONE

The interactions between radiation and the molecules in the atmosphere are not only important as a filter. The energy has to go somewhere, and when a photon is absorbed its radiation energy is converted into chemical energy and heat. When photons of wavelengths less than 240 nm hit oxygen molecules (O_2) in the stratosphere, the oxygen atoms break loose from each other. The odd oxygens are very reactive and soon combine with normal oxygen molecules forming a new compound—ozone (O_3). Most ozone is formed in the stratosphere creating an ozone layer at approximately 25–40 kilometers above the surface of the Earth (Figure 2.3).

The ozone is itself very effective in absorbing photons in the UV-B wavelengths, leaving a very different radiation environment below the stratosphere. The ozone layer probably formed about two billion years ago after blue-green bacteria had evolved and started producing oxygen. Before there was any ozone layer, all life forms had to find shelter from the UV

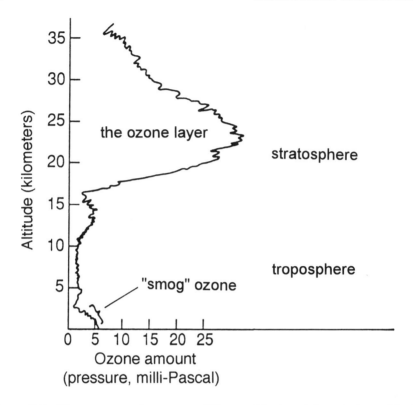

Figure 2.3   The amount of ozone at different altitudes. Adapted from *Scientific Assessment of Ozone Depletion: 1994*

radiation and it is no coincidence that life first evolved in the sea. Today, approximately one molecule out of every two million present in the atmosphere is ozone. If the ozone layer was compressed to the atmospheric pressure at the Earth's surface, it would only be 3 millimeters thick.

Ozone is not a stable molecule. The absorption of radiation of wavelengths around 320 nm breaks down the molecule and converts it back to molecular oxygen and an odd oxygen atom. Ozone can also be broken down through a series of chemical reactions, some involving oxygen and others involving odd hydrogens, odd nitrogens or chlorine atoms.

The 1995 Nobel Prize in chemistry was awarded to the three scientists who discovered the mechanisms for these chemical reactions. Paul Crutzen worked out the role of nitrogen oxides in the natural creation and destruction of ozone whereas Mario Molina and Sherwood Rowland described the role of chlorine from CFCs in these reactions (Figure 2.4).

The total concentration of ozone in the atmosphere depends on the balance between the reactions that create ozone and the ones that destroy it. Locally, the transport of ozone-rich or ozone-poor air also plays a major role.

A THINNING SHIELD

Since the 1970s, the amount of ozone in the stratosphere has been decreasing, indicating that something has upset the balance between creative and destructive forces. The trend is especially clear over the polar regions, with extremes over Antarctica.

At first the cause of ozone depletion was controversial. Among the explanations were natural meteorological phenomena, volcanic eruptions and man-made chemicals. The early suspicion that long-lived chlorinated chemicals are the major ozone destroyers is now scientifically well established. The major culprits are chlorofluorocarbons, CFCs, but the related HCFCs also play a role, as do bromine compounds used for fire-fighting, the so-called halons.

The most important evidence for the role of man-made chemicals comes from direct atmospheric measurement of some of the intermediate compounds in the destructive reactions, such as chlorine monoxide (Figure 2.5). Meteorological phenomena contribute to the day-to-day and year-to-year

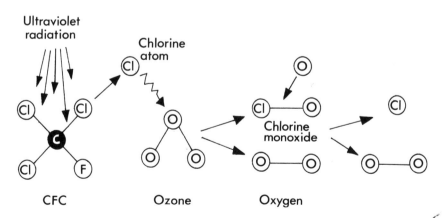

Figure 2.4   Ultraviolet radiation splits off the chlorine atoms from CFC. An odd chlorine is very reactive and immediately attacks ozone, producing chlorine monoxide and oxygen. Any free oxygen atom soon steals the oxygen from the chlorine monoxide, leaving the odd chlorine ready for another attack on ozone. Illustration: Lena Adamsson

variations in the thickness of the ozone layer, but there is nothing to show that they cause the downward trend.

How much of the ozone layer has disappeared? The thickness of the ozone column has been continuously measured by Dobson spectrophotometers from the ground for the past 50 years and via satellite since 1978. Data for 1978–89 from the Total Ozone Mapping Spectroradiometer (TOMS) on the *Nimbus 7* satellite show that the total ozone column has decreased by 3 percent between the latitudes 65° S and 65° N. The depletion varies greatly both geographically and depending on the season. Over the populated mid-latitudes (30–60°), the average rate of depletion is 4–5 percent per decade, with an accelerating rate during the 1980s (Figure 2.6). Over the Arctic, the cumulative ozone depletion is about 20 percent, while there are no significant trends in the tropics.

THE OZONE HOLE

Over Antarctica, the destruction of ozone is speeded up by a combination of factors that are unique to this area. Each winter, a polar vortex isolates the large mass of Antarctic stratosphere. There is no sun and the air becomes extremely cold. The cold, in turn, encourages the growth of ice clouds. The ice

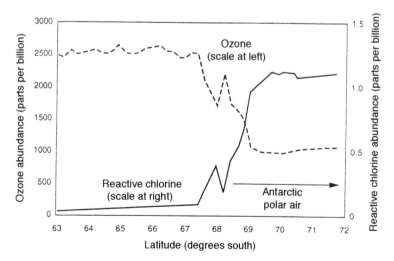

Figure 2.5   Measurements from high-flying airplanes have shown that low ozone correlates with high levels of chlorine monoxide, which is an important piece of evidence for making CFCs the culprits of ozone depletion. Source: *Scientific Assessment of Ozone Depletion: 1994*

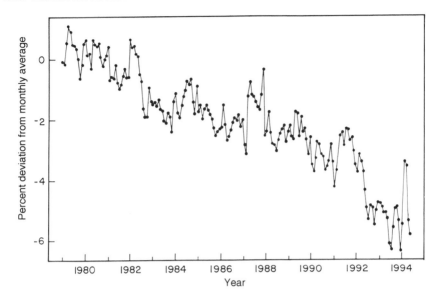

Figure 2.6   Global ozone trend 60° N to 60° S. Source: *Scientific Assessment of Ozone Depletion: 1994*

crystals in the clouds provide a surface on which the chlorine is converted to an active form when the light returns in the spring. The stratosphere is essentially primed for destruction. The extreme climate thus explains why ozone depletion is worst in this particular location. The rapid destruction stops when the polar vortex breaks up, which allows the temperature to increase and the ice crystals to evaporate. Eventually, the ozone-thin air is shuffled northward away from the south pole.

For about two months every southern spring, the total ozone declines by about 60 percent over most of Antarctica. In the core of the ozone hole, more than 75 percent of the ozone is lost and at some altitudes the ozone virtually disappeared in October 1993. In 1995, the area with severely depleted ozone was as large as all of Europe and almost twice the size compared to the two previous years (Figure 2.7). The destruction was also more rapid than previously recorded.

Recent experiments in the Arctic have shown that the mechanisms for extremely rapid ozone depletion are present here as well. Fortunately, the Arctic atmosphere generally does not get cold enough for a full-blown ozone hole to be created. Yet.

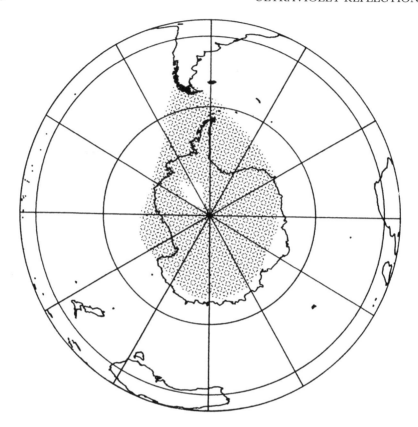

Figure 2.7   In September 1993, the ozone hole reached the tip of South America (shaded area = less than 200 Dobson Units), with total ozone values less than 100 Dobson Units over parts of the Antarctic continent (white area within shading)

THE ROLE OF VOLCANOES

From time to time, volcanoes are blamed for the input of chlorine in the stratosphere. This has been one of the main arguments by the critics of ozone depletion as a man-made phenomena. But volcanoes cannot account for the large amounts of chlorine.

This does not mean volcanoes are innocent. They do indeed have a role in ozone depletion but by a completely different mechanism. Large eruptions contribute sulfur dioxide, which is rapidly converted to sulfuric acid aerosols. The aerosols speed up the destructive chemical reactions in a way similar to the ice crystals. This can explain the severity of the ozone depletion in the years immediately following the eruption of Mount Pinatubo. However, the effects of Mount Pinatubo only lasted about two years. As a comparison, some of the man-made chlorine compounds stay in the atmosphere for more than 100 years.

GROUND-LEVEL OZONE

Ozone is not only present in the stratosphere but also closer to Earth, in the troposphere. Some of the tropospheric ozone originates from the stratosphere and has been transported downward by weather systems with the ability to move air masses over large altitudes. Ozone can also be created in the troposphere when UV radiation interacts with nitrogen oxide and other pollutants. When the situation is severe enough it is recognized as photochemical smog. Ozone is toxic to plants and a dangerous airway irritant for people, as is further described in chapter 6 "A breath of fresh air?".

The concentration of tropospheric ozone has been increasing, especially in the northern hemisphere. This ozone can, of course, also absorb UV radiation. In some regions it probably has decreased the DNA-damaging UV radiation by about 2 percent. However, the increase in tropospheric ozone cannot compensate for the global decline in stratospheric ozone. First of all, it is smaller but, more importantly, it is regional.

OZONE DEPLETION AND ULTRAVIOLET RADIATION

As ozone is a strong absorber of UV radiation, depletion of stratospheric ozone raises the question of how much the radiation environment on Earth has changed and what it will look like in the future. The first observation is that less ozone would allow more UV-B of certain wavelengths to reach through. But it is worse than that. The second consideration is that photons of shorter wavelengths are more sensitive to the concentration of ozone. This means that a loss of ozone will not only let more photons of the same wavelengths through but also open up the atmospheric window to more energetic radiation. This has important implications for the effects of ozone depletion on living systems as it means that a small change in ozone concentration can lead to large increases in the amount of damaging energy reaching Earth. With large depletions of ozone, such as over Antarctica, the levels of damaging UV-B can increase exponentially.

The physics of ozone and light is fairly well understood by now and theoretically one can calculate the increase in UV radiation at different wavelengths for every decrease in ozone. Unfortunately, understanding UV radiation is still not as straightforward as only measuring stratospheric ozone. Before looking at some of the ways to estimate and measure UV radiation, it can therefore be helpful to take a look at other parts of the atmospheric filter that are integrated in more sophisticated models. It is also useful for anyone trying to avoid a sunburn to understand the fate of damaging photons at different times and places.

THE ATMOSPHERIC FILTER

How much radiation of different wavelengths that actually reaches the surface of the Earth depends both on the composition of the atmosphere and on how far the light has to travel through the atmosphere. Besides oxygen and ozone, the photons will encounter water drops, ice crystals in clouds and aerosols formed from dust and industrial emissions, as well as sulfate aerosols from the burning of oil and coal.

Clouds are easy to see but difficult to take into account in models as the thickness and optical properties are at least as important as the amount of cloud cover. A thin cloud cover or occasional clouds in the sky might not restrict the fate of photons at all. You get almost as burned as if the sun had been shining from a clear sky. A thicker cloud cover with the right optical properties can, on the other hand, locally eliminate any effects of ozone depletion.

Aerosols, as it turns out, play a major role in shielding industrialized areas from the increases in UV radiation caused by ozone depletion, especially in the northern hemisphere. The decrease in UV-B is somewhere between 6 and 18 percent in polluted regions.

The magnitude of the effects of the different gas molecules and particles depends on how many of them the light of different wavelengths encounters on the way through the atmosphere. The yellow-red hue of the sky at sunset shows how the shorter, bluish wavelengths are filtered out when the sun is low on the horizon and the light has far to travel. At noon, on the other hand, the sun is high in the sky and the radiation has a shorter distance to travel through the atmosphere. In this case more radiation of shorter wavelengths reaches through the atmosphere, including UV-A and UV-B. In the summer between 20 and 30 percent of the total daily UV-B radiation is transmitted in the two hours around noon and about 75 percent of the total daily UV-B radiation is transmitted between nine in the morning and three in the afternoon. It is during the hours around noon that the risks of getting a sunburn are the greatest.

The local UV irradiation also correlates with latitude. The closer to the equator, the higher the sun is in the sky, effectively increasing the amount of UV radiation that penetrates all the way through the atmosphere. Further away from the equator, the sun angle depends on the time of year, with extreme variations close to the poles, where the sun does not even make it above the horizon around the winter solstice. Around midsummer, on the other hand, the days are not only long, but the sun reaches higher in the sky. The differences are noticeable when looking at the effects of sun on untanned skin. It is easy to get a tan or sunburn around midsummer, whereas the winter sun hardly leaves any mark at all. The characteristic reddening of a sunburn can be quantified in something called minimal erythema dose (MED), which is a good way to illustrate the importance of latitude. At 50° N (Belgium), the burning capacity of the sun in the fall and spring is 3 MED during one day, which can be compared to 20 MED on Hawaii at 20° N.

Altitude can also make a difference. The radiation does not have to travel as far to reach a high mountain top compared to the valley, and the probability of photons being absorbed or scattered back in space decreases. Every 300-meter climb increases the skin-burning effectiveness of the sun by 4 percent. The effect is even more noticeable in the winter when the light is reflected in the snow, which is why it is so easy to get sunburned when skiing.

IS ULTRAVIOLET RADIATION INCREASING?

With all the seasonal, geographical and weather variables in the back of our minds, we can go back to the question of how ozone depletion has changed the radiation environment on Earth. What can we see today?

To estimate the trends in UV radiation rather than ozone concentration is a tricky business. There are two different strategies, which often complement each other. The first is through mathematical calculations in computer-based radiative transfer models. These models calculate the radiative energy reaching the surface of the Earth based on solar output, solar angle and measured concentrations of compounds that absorb or scatter photons of different wavelengths. The ozone measurement of course plays an important role as input in the calculations. The models provide a theoretical map, which can be compared over time as ozone levels decrease. A major obstacle and source of error in these calculations has been the effects of aerosols and ground-level ozone over industrialized regions. These have been difficult to take into account until very recently and could explain some of the discrepancies between calculated increases in UV and what has actually been measured. Clouds also pose a problem for the modelers as they often lack good input data.

The other approach is to do direct measurements of UV radiation. Networks for direct measurements of UV radiation were first established in the early 1970s. The instruments, Robertson–Berger (RB) meters, weight the spectral qualities of the light in such a way that they measure erythema doses, which is the capacity of the radiation to cause a sunburn. They can thus provide some data for human health studies. The disadvantages in using RB measurements to look at trends are that the instruments have not been very well calibrated and that the global coverage of the network is sparse. Even locally, the measurements have been difficult to use for looking at long-term UV increases caused by ozone depletion. The UV radiation has actually decreased according to some measurements. Some of this might be caused by bad calibrations. The shielding effect of pollution can also contribute. Another problem with these instruments is that they do not give any detailed information on the energy levels in different parts of the spectrum. This makes the values less usable for assessing effects on biological systems that react differently than human skin.

Routine measurements with detail on the spectral qualities and energies at different wavelengths started fairly recently and are only made in a limited number of locations. Since the beginning of the 1990s there are about 50 instruments around the world. These instruments give much more detailed information, but the lack of old data is a serious drawback in looking at trends. And the problem is difficult to rectify. Even if a fine network of similar instruments was set up, there is no base-line of unperturbed atmosphere with which to compare the results. The measurements can thus only give a picture of the changes in UV radiation from the time when the monitoring started.

The direct measurements of UV radiation show increases in areas with severe ozone depletion, such as Antarctica, New Zealand and Australia. At Palmer Station on the Antarctic Peninsula, the levels during the springtime ozone hole are higher than summertime values in San Diego, California. In other areas the changes are small and it is difficult to draw any conclusions about trends. The first persistent increase over populated areas caused by ozone depletion was measured in 1992/93, when ozone depletion was extra severe because of the eruption of Mount Pinatubo. A Canadian measurement showing a 35 percent increase in UV-B in 1993 compared to the four previous years has received a lot of attention, but there are some problems with this number in looking at long-term changes. The UV radiation is now back to 1992 levels, and the extreme increase was probably not representative of the general trend.

By combining the direct measurements with results from radiation transfer models, it is possible to get a more complete picture of trends in UV radiation. Over the years 1979–89, the UV dose, weighted for DNA damage, increased by about 10 percent at northern high latitudes, by 5 percent at 30° N and 30° S, 15 percent at 55° S and 40 percent at 85° S. In the tropics there has not been any significant trend. These numbers do not take the effect of cloudiness and local pollution into account.

ULTRAVIOLET FORECASTS

The regional estimates of UV radiation will not say much about the local situation on a particular day. That will depend much more on small-scale changes in ozone as well as the weather and how high the sun is in the sky. Therefore, several countries are setting up UV indexes as a service to people who want to avoid getting too much sun. They are typically based on a forecast of ozone levels, weather, cloud cover, time of year, and how reflective the ground is. The spectral qualities are weighted to indicate the damaging effect on skin.

Different national systems are now being harmonized and in the future everyone will use a scale from 0 to 15. Zero indicates no UV whereas any value over 10 is very high. Table 2.1 is an example from Sweden showing how the UV index can be interpreted for someone with a skin type that always turns red and sometimes tanned. People with more sensitive skin will have to reduce their time in the sun even more.

Table 2.1   The UV index gives advice about how long it is safe to stay in the sun without getting sunburn

Index		Typical of	Time in sun before skin turns pink
2–4	Low		1–2 hours
4–7	Moderate	Northern European summer	30–60 minutes
7–10	High	Summers in southern Europe and on Canary Islands during spring fall	15–30 minutes
10+	Very high	Canary Islands during the summer and Equatorial areas	10 minutes

ULTRAVIOLET RADIATION IN THE FUTURE

What will the future UV-radiation environment look like? The answers lie as much in politics as in atmospheric science. There is an international consensus to phase out most ozone-destroying chemicals, and if the commitments are met, chlorine levels will only increase up to a certain point. There is, however, a big question mark about how well some developing countries and countries in former Eastern Europe will be able to live up to these goals. Developing countries have a 10-year grace period after the international phase-out deadlines, and there is a certain risk that their use of ozone-destroying chemicals will increase drastically during this time.

If everyone follows the international agreements in the Montreal Protocol, the amount of chlorine in the atmosphere will peak around 1998. Using model simulations, it is possible to predict that the ozone depletion caused by this chlorine will peak in the next few years. But the chlorine compounds have a long lifetime in the atmosphere and it will be at least half a century before the ozone layer recovers completely. Compared to 1960, the worst ozone depletion at northern mid-latitudes will be 12–13 percent in the winter and spring and 6–7 percent in the summer and fall. The depletion at southern mid-latitudes will be about 11 percent.

Is this decrease in ozone small or large? Compared to daily and seasonal fluctuations, it is not very alarming. It would be of the same magnitude as the difference between a cloudy and a sunny day. However, the total UV dose will of course increase as the ozone depletion will be superimposed on the normal fluctuations (Figure 2.8). Moreover, the peak doses will be higher than that which living systems have had to adapt to previously.

TRANSLATING TO BIOLOGICAL EFFECTS

The answer to questions about how big a change we are in for will also depend on how each biological system reacts. Is the increase in damage a linear function of the decrease in ozone or could the detrimental effect double or even triple for each percent decrease? The rest of the book will discuss in detail some of the problems in predicting the future. However, there is a term often used in risk assessments that can give an indication of the magnitude of change: radiation amplification factor (RAF). RAFs take into account the specific wavelengths to which a biological system is sensitive and how ozone depletion affects those wavelengths. For example, the DNA molecule is very sensitive to UV at wavelengths shorter than 310 nm. The drop-off in response

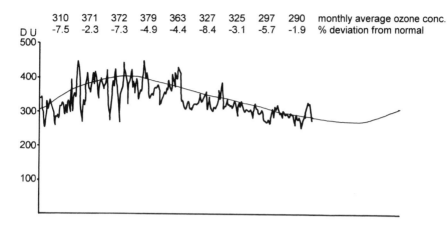

| 310 | 371 | 372 | 379 | 363 | 327 | 325 | 297 | 290 | monthly average ozone conc. |
| -7.5 | -2.3 | -7.3 | -4.9 | -4.4 | -8.4 | -3.1 | -5.7 | -1.9 | % deviation from normal |

JAN FEB MAR APR MAY JUN JUL AUG SEP OCT NOV DEC

Figure 2.8   Ozone measurement from Norrköping, Sweden, show that, in spite of large daily variation, total ozone was below average for the first nine months of 1995. Source: Weine Josefsson, SMHI/NV Environmental Monitoring Data

is sharp and the molecule is hardly affected by wavelengths longer than 320 nm. In such a case the effects of ozone depletion can be quite pronounced as the decrease in ozone opens up a window for exactly those wavelengths to which DNA is sensitive. A typical RAF for DNA damage is two, implying that each percent decrease in ozone gives a 2 percent increase in DNA damage. Reddening of the skin in a sunburn can also be caused by radiation of longer wavelengths, even in the UV-A band. It would thus be relatively less sensitive to ozone depletion, which is reflected in a lower RAF (Figures 2.9 and 2.10).

Using RAF values, increases in damage-weighted UV can be calculated based on the expected peak ozone depletion (see Table 2.2). The estimates warrant a word of caution. First of all, they depend on everyone following the international agreements to phase-out ozone-damaging substances. Moreover, they do not include any atmospheric surprises leading to more ozone depletion than the present models suggest. The numbers are also difficult to use when looking at the damage to complex biological systems, where other mechanisms can reduce or enhance the direct effects of UV radiation. For example, the RAF for DNA is based on the effects of the naked molecule, rather than on the situation inside a cell.

For areas with sharp declines in ozone, such as Antarctica, there are special problems using RAFs. Most of the values in the literature only hold up for small decreases in ozone. Recent observations based on direct measurements

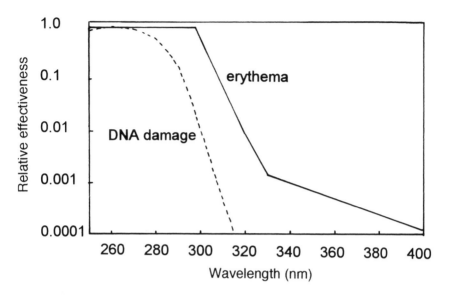

Figure 2.9   Action spectra for DNA damage and erythema showing that erythema is relatively more sensitive to longer wavelengths than DNA. Source: *UV Radiation from Sunlight*

of UV radiation in Antarctica show that the response of UV radiation to ozone depletion is not linear. Rather, with severe ozone depletion, the amount of UV radiation increases exponentially, which creates completely different RAF values. For example, a 50 percent decrease in ozone can give a 350 percent increase in the DNA damage.

A GLOBAL EXPERIMENT

Another way to discuss the magnitude of change in UV radiation is to compare the projected levels in an evolutionary perspective. Will the radiation be harsher than life has previously experienced? Unfortunately, no one really knows as there are no geological or older historical records of the ozone concentration in the atmosphere. What we do know is that many life forms, especially plants, constantly try to balance the positive effects of the sun with its negative impacts. The assessment of the effects of ozone depletion thus becomes a question of how the positive and negative aspects of solar radiation will balance each other. A main concern is that thresholds may be reached that will tip something or someone over the edge—a population of plankton,

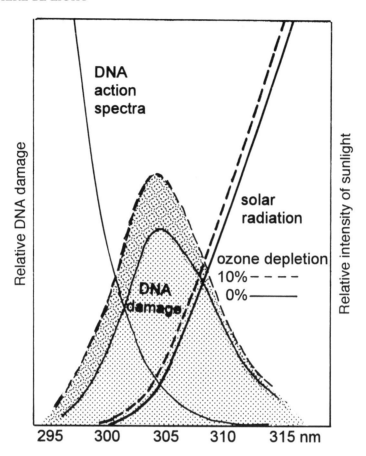

Figure 2.10   The biological effectiveness of ozone depletion can be illustrated by looking at the intensity of solar radiation under different ozone levels. The shaded areas under the curves are proportional to the effect of the radiation on DNA. Adapted from *Strålning. Energi i rörelse*

Table 2.2   The effect of ozone depletion depends on the sensitivity of each biological system. The following percentage increases in erythema and in DNA damage can be calculated from the expected peak ozone depletion

Damage	Erythema	DNA damage
Northern hemisphere, winter/spring	15–17	29–32
Northern hemisphere, summer/fall	8–9	12–15
Southern hemisphere, all seasons	15	25

the immune system of a human being or even a biogeochemical cycle governing the climate of this planet. We are in the middle of a UV-radiation experiment on a global scale. Trying to mimic the future can give us a glimpse of what to expect.

Chapter 3

Plankton life under the ozone hole

WE ARE LOOKING FOR GREEN WATER. THE WATER sampler has just been brought aboard the American research vessel *Nathaniel B. Palmer* and everyone is busy setting up for filtering for phytoplankton and analyzing chlorophyll. In the background, an optical instrument constantly records the color of the water we are passing through.

At the surface, the water appears gray under the overcast sky, but the instruments give a different picture. Soon the scientists on board have found what they want—an area where the sea is green with plankton. It is like a spring meadow in full bloom. A look into a bucket of water confirms that the sea is full of sun-catching cells. Most prevalent is a species of colony-forming diatom. Under the microscope, the individual cells in the colony look like beads stuck in a mass of jelly. We are about 100 kilometers from the edge of the sea ice: 59° 35' S and 49° 59' W at the Weddell–Scotia Sea in the Southern Ocean (Figure 3.1).

Soon the water sampling is daily routine, if anything can be called routine when working in one of the stormiest seas in the world. Wind and waves batter the equipment when it is brought up from 200 meters depth on board the rolling ship. Safety lines and life-jackets are standard dress code when the door on the side of the ship opens up (Figure 3.2).

At the same time, various experimental designs find their way up to the helicopter deck of the vessel. There is a simple version with glass tubes in a water-filled wooden crate and the more intricate model with rotating screens to simulate the movement of plankton in the water column at various depths of the sea. The purpose is similar—to look at how the plankton fare under different light conditions. The goal is to understand how the ozone hole affects the microscopic plants that form the base of the Antarctic food chain.

A disturbance in the plankton community has two important implications. One is that there will be less food for the organisms that feed on these tiny plants, such as krill. The krill, in turn, are the major food source for birds and whales in the Southern Ocean (Figure 3.3). The role of the krill for wildlife is clear from the Cape pigeons that constantly follow the ship to get at the tiny shrimp-like creatures the propellers whirl up to the surface. Every once in a while the krill also attract chin-strap penguins and smaller whales. Or maybe they are just curious about the strange noisy creature that breaks the peace in their home waters.

The other reason to worry about the fate of phytoplankton under the ozone hole is the role they play in the global circulation of carbon, which in turn affects the concentration of the greenhouse gas carbon dioxide in the atmosphere. For the global climate, the prolific plankton blooms in the

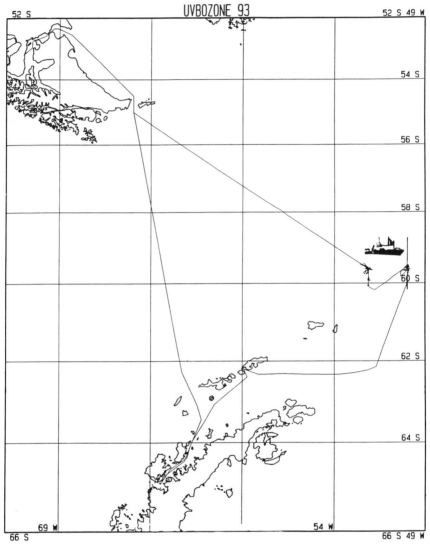

Figure 3.1   The voyage with the *Nathaniel B. Palmer* started in Punta Arenas, Chile, and went via the Antarctic Peninsula to the study sites at the Weddell–Scotia Confluence

Southern Ocean have the same role as the trees growing in a tropical rain forest. The problem in Antarctica is that the plankton are adapted to very low levels of ultraviolet radiation and their ability to harvest carbon might suffer because of the changes in the light environment.

Figure 3.2   Water sampling from the *Nathaniel B. Palmer*. Photo: Annika Nilsson

INCREASING LIGHT STRESS

The ozone thinning was first detected in Antarctica in 1985 and since then there has been an intense research effort into measuring UV radiation and trying to understand the implications of any change. Antartica is one of the places where the UV radiation has increased significantly in the past 10 years. This is especially true in the Antarctic spring when the ozone depletion is so severe that it can be called an ozone hole. The satellite images from the middle of October in 1993 showed that the ozone column was less than half its normal value in an area that extended as far north as 60° S. Since then even more severe losses have been recorded.

There is good reason to suspect that the severe ozone depletion has negative effects on the productivity of plant life in the water. Already under normal light, with no ozone depletion, light damage can hamper the sun-catching capacity of phytoplankton. The lack of ozone in the atmosphere allows UV light of shorter, more energy-rich wavelengths to reach the surface of the Earth. Depending on how much of the damaging radiation actually reaches the cells in the water, more inhibition can be expected.

In 1990, these suspicions were also confirmed in the field by researchers in

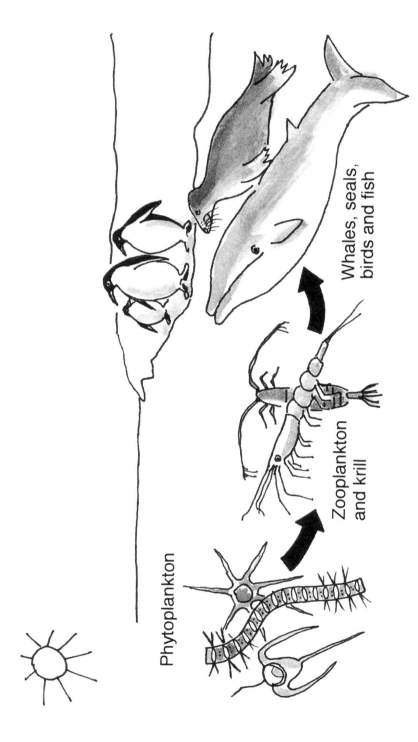

Phytoplankton

Zooplankton and krill

Whales, seals, birds and fish

Figure 3.3 The Antarctic food web. Birds, whales and fish are dependent on krill, which in turn feed on phytoplankton. Illustration: Lena Adamsson

the project Icecolors. In an elegantly designed experiment, they used the fact that the ozone hole is oval in shape and that it rotates over Antarctica. A ship staying in a suitable position can use this rotation to measure the productivity under different light conditions—inside and outside the hole. It becomes a controlled experiment in the most natural situation possible: using the native plankton community under actual solar radiation. By use of a new instrument they could even measure the amount of light that reached the plankton at the different depths they had placed their incubation chambers. Their results clearly showed that the productivity of the plankton decreased under low ozone conditions. They estimated that the total loss of primary production in the blooming areas of the Southern Ocean was somewhere between 6 and 12 percent.

Experiments with submerged containers of phytoplankton at the American research station Palmer at Anvers Island have confirmed the picture. The results show that the plankton suffer, with an estimated 4 percent loss in primary production. The two estimates from Icecolors and Palmer Station indicate that there is no consensus of how big the problem really is. Osmund Holm-Hansen at Palmer Station does not think the overall effect is very serious at all as much of the Southern Ocean is covered by ice during the most severe ozone depletion. The ice would cut out most of the damaging radiation.

SENSITIVE TARGETS

Why do the plankton slow down their sun-catching activities when exposed to UV radiation? The past years of research on plankton in cultures and in more accessible waters add up to a substantial bank of knowledge on the physiological effects of light on cells and on their dynamic response.

A well-known effect is bleaching of the sun-catching pigments. It has been shown that the concentration of all major photosynthetic pigments in natural Antarctic plankton decreases after one to two days of exposure to UV radiation. Ultraviolet light can also damage the DNA molecule, which holds the genetic information in the cells, and the RNA molecule, which is important in translating this genetic information. Both of these molecules absorb energy in the specific UV-B wavelengths and build molecular bridges that destroy their biological function. The cells have some ability to repair the damage by molecular cutting and pasting, but certain damage is more serious and cannot be repaired. Ultimately, damage to DNA and RNA will interfere with how the genetic information is read and translated and thus affect the survival of the plankton cell.

Another sensitive system is the photosynthetic machinery, the molecules that are responsible for catching light energy and transferring it to chemical bonds in the cell. Even under visible light, the structure of one of the key molecules can be damaged. Measurements of the rate of photosynthesis show that this has a direct effect on the phytoplankton's ability to harvest solar energy. The long-term consequences for plankton productivity depend on how fast the cell can repair or replace the damaged part, which requires synthesis of new proteins. As long as the protein synthesis can keep pace with the damage, the plankton will be able to photosynthesize.

There is also less specific damage done by UV radiation to the cell. For example, there is a general decrease in protein content of plankton that have been exposed to UV radiation. Not much is known about what systems in the cell are damaged, but if enzymes that are important in cell metabolism are damaged, it would clearly affect the cell's ability to grow and survive. The membrane surrounding the cell is another sensitive target as it is responsible for all transport of material in and out of the cell, including the carbon necessary for photosynthesis.

Plankton do not live on carbon and light alone. Nutrients, such as nitrogen, are important for many metabolic processes. There are some signs that the ability of plankton cells to utilize nitrogen might be even more sensitive to UV-B than their carbon-fixing ability.

How do the different kinds of damage to individual molecules in the cell add up? The answer depends on the rate of damage and also on other environmental stresses. Plankton have evolved some ability to repair damage caused by the solar rays. However, if the damage is too great, the cell will have trouble keeping up. For example, the repair requires substantial rates of protein synthesis to rebuild the photosynthetic machinery. Damage to the genetic blueprint and its translation would clearly hamper that repair.

A COMMUNITY APPROACH

A different approach to understanding plankton sensitivity to increases in UV radiation has been to look at the plankton community as a whole. The focus has been on how different light regimes affect the primary production, which is a measure of how well the plankton manage to capture light energy and bind carbon. In the late 1970s and early 1980s, it was noticed that the standard glass containers that were used for culturing plankton could protect the cells from some damage caused by light. It soon became clear that the glass cut out light of shorter, more energetic wavelengths. Further studies have since shown that long-term exposure to UV-A and UV-B clearly hampers photosynthesis in the plankton cultures.

At first these experiments were only carried out in laboratory cultures under artificial light sources. During the past few years they have increasingly been done under more natural conditions, such as the Icecolors cruise and research at Palmer Station. Measuring the primary production—the ability to bind carbon—gives a sum of all the positive and negative effects of light on the plankton community. On the plus side is the fact that light provides energy for photosynthesis and that it serves as a signal for different repair systems and protective responses. On the minus side are the different kinds of damage: to DNA and to the photosynthetic machinery, as well as the energy costs of producing protective pigments and repairing damage.

ACTION SPECTRA

A further sophistication of looking at primary productivity is to gain a systematic knowledge of how sensitive plankton are to light of various wavelengths and to create a mathematical model where it is possible to plot the light conditions versus primary productivity. This is called an action spectra. It is essentially an equation which describes the magnitude of the effect on phytoplankton for each incremental increase in UV radiation. If it is properly determined, the equation should give a tool to estimate the effects on the plankton population of a known increase in UV radiation.

Action spectra differ for different biological processes, depending on which wavelengths are the most damaging. In modeling the effects of UV radiation on living systems, it is necessary to determine the response to doses at different wavelengths for each system. Action spectra for plankton are fairly new and pet projects for two of the scientists on board the *Nathanial B. Palmer*: Patrick Neale and John Cullen. Their goal was to verify an equation that had been determined on plankton in tanks and to compare the plankton in the marginal ice zone in the Weddell–Scotia Sea with other plankton communities (Figure 3.4). The basic procedure was to measure the photosynthetic rate as the uptake of radioactively labeled carbon in small vials with plankton-containing sea water. The water is exposed to a well-defined light source, strong enough to imitate the sun, and by varying the intensity of light with screens and the wavelength with filters, one can determine the net result of different light regimes on the photosynthetic rate of the plankton.

It was long hours of work in a small and cold container on the helicopter deck, keeping track of vials, carbon-14 and exposure times. The equipment

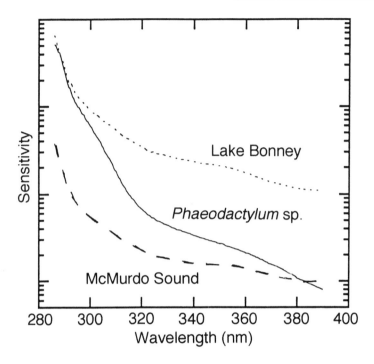

Figure 3.4   Previous studies of phytoplankton from Antarctic environments (McMurdo Sound and Lake Bonney) and for a temperate diatom (*Phaeodactylum*) show that photosynthesis is sensitive to UV radiation and that sensitivity increases sharply in the UV-B range affected by ozone depletion. Source: *Ultraviolet Radiation in Antarctica*

had to be securely tied down or fastened with duct tape to deal with the almost constantly rolling sea. The effort paid off, but the results were a surprise: the short-term response of the Antarctic phytoplankton was fundamentally different from other plankton. The inhibition of photosynthesis seemed to be irreversible for serveral hours. It was as if the plankton could not repair the damage. It means that one cannot assume a balance between damage and repair in the models. This also confirms the suspicion that Antarctic marine life is adapted to very low light conditions, making any change in radiation environment potentially more devastating here than elsewhere.

The next step is to translate the results into the reality of the Southern Ocean, where currents and wind enter the equation. The lessons from the Weddell–Scotia Sea clearly show that in the ocean plankton do not sit still waiting to be illuminated by a constant light source.

STORM IN THE SPRING MEADOW

White-capped waves and constant high swells as the gale beats the water surface is the manner in which the Weddell–Scotia Sea greets its visitors in the spring. Gales and storms seem to be the normal conditions. Each wave sets up a mixing current, which adds to the other forces that move plankton around in the water column. The result is a mixed layer of 80–100 meters in depth (Figure 3.5). For the plankton this means they might only spend a short time at the surface, where they can be hit by the damaging UV light.

Sunshine was a rare occurrence during the spring cruise in the Weddell–Scotia Sea, but one should not be fooled by gray skies. The light sensor in the top of the highest mast showed elevated levels of UV radiation each time the ozone hole was positioned over the ship. The levels were such that one would expect some damage to the plankton community if it was exposed long enough. One key question is how exposed they actually were. The sea water absorbs much of the radiation, but differently depending on the wavelength—the shorter the wavelength, the better the radiation is absorbed. The result is that visible light penetrates much deeper into the water column than UV light, UV-A penetrates deeper than UV-B, etc. The underwater optical instruments that we winched down in connection with every water sample showed that most of the UV-B was typically absorbed in the top 5–10 meters. Photosynthetically active radiation penetrated down to at least 30 meters (Figure 3.6). These numbers are by no means a general description of the light regimes in the Southern Ocean. The light penetration depends on the amount of plankton and other particles in the water. Measurements from Icecolors 90 show that UV-B can reach as deep as 60–70 meters even if there is too little to inhibit photosynthesis. Inhibition has been detected as deep as 25 meters.

The stormy reality of the Southern Ocean makes it very clear that to get an accurate picture of what happens in the ocean, one also has to enter time into the equation. How long are the plankton exposed? How fast are they swept away from the surface by currents before being hampered? The measurements on board the *Nathaniel B. Palmer* showed that a one-hour exposure to light that is similar to the sun under the ozone hole is enough to completely stop all photosynthetic activity of the plankton. If the plankton only had half-an-hour exposure, they retained some photosynthetic activity, but were slow to recover. With the waves and water mixing patterns during the cruise, this is important information as a single plankton cell is very unlikely to stay at the surface for very long at all.

The role of mixing was further confirmed by one of the experimental set-ups on the helicopter deck. By simulating the movement of plankton to

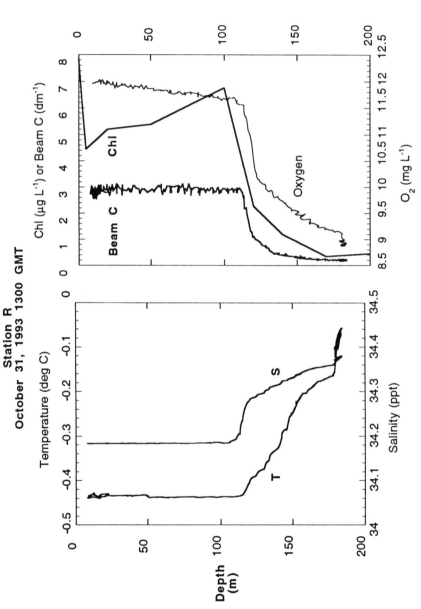

Figure 3.5 The mixing layer in the Weddell Sea was very deep, as indicated by measurement of temperature, salinity, chlorophyll (Chl), oxygen (O_2) and turbidity (Beam C) of the water

Figure 3.6   Getting ready to measure the penetration of UV into the water on one of
the rare sunny days during the cruise in Antarctica. Photo: Annika Nilsson

various depths with the use of rotating screens cutting out some of the light,
Gustavo Ferreyra and Irene Schloss could show that the mixing could indeed
lessen the damaging effects of UV radiation under the ozone hole (Figure 3.7).

The role of mixing might not always be positive, however. There is a risk
that it can instead increase the damage as the plankton community is more
adapted to low light conditions deep in the water than to the sun-lit surface.
Each plankton can adjust to strong light by producing sun-screening
pigments, but this adjustment takes a while. If the plankton never remain long
enough at the surface, they might never get time to produce enough pigment.
John Cullen describes this as a "conveyer belt of doom"—the water constantly
bringing up light-sensitive plankton to the strongly illuminated surface water.

The role of plankton movement in increasing or decreasing the damage
caused by ozone depletion makes it clear that to get an estimate of the
productivity in the ocean, the light measurements and the information from the
action spectrum have to be complemented with systematic studies of mixing
rates. It also calls into question the value of productivity measurements that
have been done by illuminating a sample at one specific depth for a whole day.

A third problem is extrapolating productivity measurements from one area

Figure 3.7   This arrangement on the helicopter deck of the *Nathaniel B. Palmer* was used to simulate the movement of plankton in the water. Photo: Annika Nilsson

of the Southern Ocean to another. How representative was the roughness of the sea and deep mixing layers in the Weddell–Scotia Sea? What is the light environment for plankton in more protected areas and if the weather is calm for an extended period? The plankton-containing mixing layer could be much shallower, which would restrict the vertical movement and potential for protection from the most intensive UV radiation.

    Moreover, how representative is one particular plankton community for the Southern Ocean as a whole? The studies on the *Nathaniel B. Palmer* showed that the sensitivity of the plankton varies considerably, both in time and space. In fact, this variable sensitivity has a much greater influence on calculations of water-column inhibition of photosynthesis than the full range of stratospheric ozone values encountered during the Antarctic spring of 1993. At one of the stations, a shift in ozone values from 320 to 140 Dobson Units caused a decline of 3 percent in water-column productivity, while the ozone depletion caused no change at all in the productivity at another station. The overall effect of UV-A and UV-B on the plankton ranged from 7.5 percent to 18 percent of daily production.

With all the unknowns and the natural variability, it is not so strange that some people see large threats to Antarctic plankton productivity where others see only minor problems with ozone depletion in this particular respect. All estimates of primary productivity of the Southern Ocean under the ozone hole are educated guesses based on slightly different experimental designs and slightly different assumptions.

FOOD FOR THOUGHT

Looking at rates of photosynthesis is only part of the description of the effect of increases in UV-B over the Southern Ocean. Even if the direct loss of photosynthesis is small, there might be shifts in the structure of the plankton community, which in turn have ecological consequences higher up in the food chain. For example, in the austral spring of 1993, the Weddell–Scotia Sea was full of a colony-forming species that made up as much as 60 percent of the biomass. Are these colony-forming plankton as palatable a food source as individual cells? It might very well be that they are too big a mouthful for many of the organisms in the next level of the food chain. Also, there are indications that the colonies are less sensitive to UV-B than the free-swimming cells. Thus, an increase in the UV-B flux might favor a plankton community structure that does not provide an adequate base for the traditional food chain of plankton–krill–mammals/birds.

Was the "clumpiness" of the plankton community indeed an adaptation to increased UV-B flux under the ozone hole? Unfortunately, not enough is known about the marine microlife previous to the ozone hole to answer that question. However, if the plankton community has the ability to adapt to low ozone conditions, that adaptation would certainly be part of what can now be seen in the Southern Ocean. The natural base-line was lost about a decade ago.

THE CHEMISTRY OF SEA WATER AND LIGHT

The increase in UV also has effects on the environment in which the plankton live by changing the chemistry of the water. Specifically, the high-energy radiation can create reactive chemical compounds, so-called free radicals. These can in turn break up complex carbon compounds in the water making more carbon available as food for the bacterial microlife of the ocean.

The reactive free radicals can also interact with metal complexes to release ions such as copper or iron. There has been speculation that iron is a limiting factor for plankton productivity in the Southern Ocean and if more iron is freed by photochemical reactions it could very well be beneficial for marine life.

The reactive chemistry of sea water and light can also have some detrimental effects on plankton life. As the light hits the water molecules, a reactive hydroxyl radical can be formed. These radicals are highly indiscriminate and interact with any biomolecule in sight, including the surface of living cells, causing chemical and structural transformations. A $1 \, \mu m$ plankton cell will experience about ten collisions with hydroxyl radicals per second in sunlight-illuminated water. Even if only a small percentage of these collisions result in degradation of a vital cell surface, it still represents an important environmental stress on plankton organisms.

In addition, reactions of the radicals with other constituents of the sea water can produce a cascade of reactive secondary radicals such as dibromide ions and carbonate. In response to this, the organism might produce slimes to protect its surface. Thus, energy normally required for growth and reproduction would be diverted to protect the cell surface with lowered productivity as the end result.

AN INTERDISCIPLINARY EFFORT

The studies of the chemistry of increases in UV in the Southern Ocean have only recently started and it is too early to draw any conclusions about the effects on plankton life. What did become clear from the work of Kenneth Mopper, David Kieber and the other chemists on board *Nathaniel B. Palmer* was that there is enough UV radiation to cause a change in photochemistry of the water under the ozone hole. The experiences from the Weddell–Scotia Sea also show that understanding the effects of the ozone hole on marine life has to be an interdisciplinary effort. Marine biologists need to cooperate with physical oceanographers to understand movement and mixing. Accurate light measurements require new, sophisticated instrumentation. Knowledge of chemistry becomes vital in understanding both the water environment and the light-induced reactions inside the cell.

Chapter 4

Sun catchers

E VENING WORK AT ABISKO SCIENTIFIC RESEARCH station in the mountains of northern Sweden: . . . 1.67 milligram, 1.15 milligram, 2.19 milligram . . . One at a time, doctoral student Ulf Johanson weighs this year's shoots from the small perennial crowberry bush *Empetrum hermaphroditum.* Together with bilberries, cowberries and bog whortleberries, the crowberries cover the ground outside the station in the low alpine birch forest. In between the scattered trees and low brush, there are mostly mosses and lichens, and downstairs at the research station Carola Gehrke is busy extracting chlorophyll from moss shoots. She grinds the shoots in acetone and evaluates the green liquid in a spectrophotometer to look at how much chlorophyll the moss has produced during the summer.

It is only the end of August, but in a few weeks the birch trees will change color and snow will cover the ground. The days are growing shorter and the research students are busy finishing up before they have to pack their equipment and take off for their home institutions at Lund University in southern Sweden.

It has been a fruitful season and, even if their results are yet to be analyzed in detail, there are some clear trends—increased ultraviolet radiation does affect the Arctic plant life. There are changes in leaf thickness and the UV-irradiated plant material is slower to be decomposed by the natural microbial community. One of the mosses, on the other hand, seems to be growing faster when irradiated with extra UV-B (Figure 4.1). Is this a temporary increase as a reaction to stress? What else could be going on? Next year might bring some answers, or the year after next. Just before the snow melts in the spring, the two students will be back to set up their lamps in the terrain. It will be another season of research into the ecological effects of increased UV radiation on what might be a very sensitive ecosystem—one that has evolved with the sun always low in the sky and thus very little natural UV-B radiation.

LIGHT AS FRIEND AND FOE

The life of a plant revolves around catching energy from the sun and converting it into chemical energy in the photosynthetic process. Light is life and there are pigments as well as intricate chemical pathways to ensure that the process works as efficiently as possible. But it is a balancing act. Even the UV radiation present in today's natural light can wreak havoc with parts of the system and decrease its efficiency. The photosynthetic machinery in itself is sensitive. So is the hereditary material of the plant cells along with some structural parts of the cells.

Figure 4.1 Measuring growth to determine whether the moss is harmed by increased UV radiation. Photo: Annika Nilsson

The early warnings about ozone depletion created a demand for knowledge about what actually happens in a plant when UV radiation hits the leaves. A number of greenhouse studies were started where plants of different species were illuminated with UV lamps. There were warnings about decreased productivity and fears that ozone depletion might have severe effects on the yields of agricultural crops. Recently, these warnings have been played down, even by international experts in the field. The reason is that the lamps did not accurately simulate the kind of changes in light that would be expected under a thinning ozone layer and new experimental designs in the field have added more accurate data. However, the greenhouse studies, along with laboratory experiments, have increased our understanding of light as a friend and a foe for plants.

The rays of the sun first hit the epidermis of the plant leaf (Figure 4.2). By inserting tiny fiberoptic devices in the leaves, it has been possible to show that the epidermal cells along with protective pigments are very effective in reducing the transmission of UV-B. Photosynthetically active radiation (PAR = 400–700 nm) and UV-A, on the other hand, penetrates fairly deep into the leaf, where it reaches the photosynthetic apparatus of the chloroplasts in the mesophyll layer.

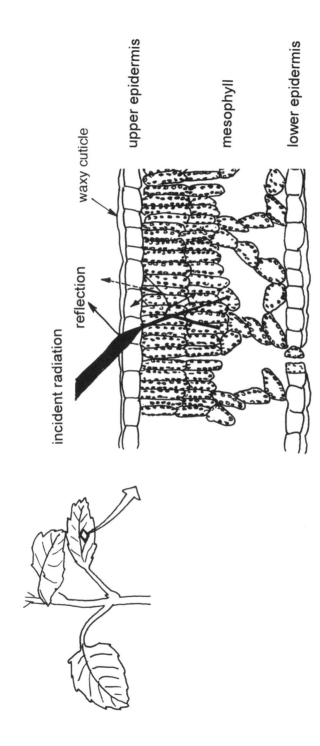

Figure 4.2 Most of the radiation is absorbed in the waxy cuticle of the leaf. Adapted from *UV Radiation from Sunlight*

However, even with intricate screening mechanisms, some UV-B will reach sensitive targets within the leaf. The most important damage is probably to the DNA. Not much research on this has been done in higher plants, but studies on bacteria and single-celled eucaryotes have clearly shown how UV-B has a direct effect on the hereditary material in the cells. There is one plant study that shows how the DNA molecule locks itself by forming bridges between different subgroups, so-called dimers.

What role does DNA damage play for the plants as a whole? The answer depends on whether the cell can repair the damage, and here lies a clue to some of the difficulties with the early greenhouse studies. All cells have elaborate systems for DNA repair, one of which is activated by light. It is an enzyme called photolyase that is activated by UV-A and by blue light. This photoreactivation is not as efficient in low-light conditions, which might explain why plant growth was so severely hampered in the early greenhouse experiments.

Another sensitive target is the photosynthetic machinery. In the laboratory, it is possible to isolate the parts of the cells responsible for photosynthesis, the chloroplasts. If these are irradiated with UV-B, there are clear hampering effects on photosynthesis. There are several possible explanations. One of the most important is the negative effect on photosystem II, which is an important light-capturing structure within the chloroplast. A second explanation is connected to DNA. If the DNA is damaged, proteins will not be synthesized fast enough to replace the damaged enzymes.

Unfortunately, neither the observations nor the molecular explanations are very useful in making ozone-depletion risk assessments. The problem is that the rate of photosynthesis does not relate directly to plant growth. Moreover, the light hitting the isolated chloroplasts in the laboratory does not simulate the light hitting the chloroplasts inside a leaf. It is thus impossible to translate the activity measured in isolated chloroplasts to what happens with productivity in an intact plant.

The third target is the cell membrane. Ultraviolet radiation makes the membrane permeable by damaging a molecular pump responsible for the transportation of potassium. This might explain an observed effect on the stomata, which regulate the uptake of carbon dioxide into the leaf. Again, it is questionable whether this has any effects on the plant as a whole as the stomata are on the underside of the leaf and thus not very exposed to the sun.

A fourth sensitive system is the cell's skeleton. It is composed of a filamentous network of microtubules and microfilaments and is important for cell growth and morphology. Research on animal cells shows how the microtubules disassemble and have trouble reassembling under UV-B radiation.

PROTECTION AGAINST HARMFUL RAYS

Even without ozone depletion, some UV-B radiation always reaches the Earth's surface, especially close to the equator and around the noon hours. Plants cannot move to keep away from the sun when it is harmful, and in addition to repair systems they have instead evolved a variety of protective mechanisms.

Some plants have a leaf anatomy that protects against too much light. For example, UV-B penetrated much deeper in herbaceous dicotyledons than in coniferous needles. Woody dicotyledons and grasses fall somewhere in between.

There are also differences in plant behavior, where some species can orient themselves toward or away from the sun, to provide the photosynthetic machinery with the optimum environment.

Some of the protective mechanisms are specifically induced by UV radiation, such as the production of flavonoids. These are colorless pigments that effectively absorb in the UV region of the solar spectrum. The flavonoids are present in leaves, pollen, petals, stem and bark and are found mostly in the epidermis.

The reproductive organs, such as anthers, as well as pollen, have extremely high concentrations of these pigments. For example, 98 percent of all UV-B is absorbed in the outer layer of the anther, effectively protecting the DNA. In pollen, 80 percent of UV-B can be similarly screened.

Another protective mechanism is the production of a waxy layer on the leaf surface. The wax does not absorb UV-B but may reflect 10–20 percent of the incoming radiation. Both the amount of wax and its chemical composition can change as an effect of UV stress on the plant.

Stress from light or from other sources can also cause the plant to produce phenolic compounds and alkaloids that protect against pests. UV-B also stimulates the production of lignin. These compounds are not specific in providing protection against UV radiation but are rather general stress signals that serve to keep herbivores away.

PALATABILITY AND COMPETITIVE BALANCE

The protective mechanisms are interesting in two respects. First, they would clearly influence the plant's ability to cope with ozone depletion and undermine any risk assessment that is based only on the molecular mechanisms of UV damage. The second, and very important aspect, is that they could change the ecological role of a plant. For example, flavonoids, other stress metabolites and lignin content will change the palatability of the plants to herbivores of different kinds.This includes the ability of microorganisms to break down plant material. Returning to Abisko and the low dwarf shrubs, it has been shown that leaves that had been exposed to extra UV-B did not decompose as quickly and that the microbial activity was lower than for untreated leaves. Analyzing the chemical composition of the leaves, it was clear that the treated leaves were richer in tannins and poorer in cellulose than the untreated leaves.

The changes in plant structure are also important in assessing the ecological effects of ozone depletion. Different plants will have different responses and changes in the anatomy of the plant might alter the competitive balance between two species. This is especially true if the changes occur in the shoot, when the plants compete for light and space, such as in a newly sown agricultural field. Wheat has been shown to gain in competition against the weed wild oat because it is less reduced in size and shades its competitor. An array of experiments looking at competitive balance between species exposed to increased UV-B revealed changes in four out of ten species pairs.

A conclusion that has been drawn from these kinds of observations is that ozone depletion might alter the competitive balance between plants even if it does not decrease the total productivity of an ecosystem. There will always be plants that are fairly UV tolerant and that will be able to take over the ecological space left by more vulnerable species.

PRIMARY PRODUCTIVITY

Despite the current focus of plant research on ecological effects and competitive balance, there is some interest in estimating the productivity of different plants under increased UV-B. In man-made ecosystems, such as tree plantations and agricultural fields, productivity will determine the crop yield and the economic profit.

The discussion on productivity starts with photosynthesis and carbon dioxide assimilation. It is the same basic process that marine researchers study in phytoplankton and it is a matter of measuring carbon dioxide uptake under different light conditions. Most studies, both in greenhouse experiments and in the field, show that net photosynthesis is decreased under a simulated ozone depletion. It is also clear that the changes are smaller than the effects on some of the individual reactions in the photosynthetic pathway. As previously mentioned, it is also clear that the negative effect of UV-B is much worse in low natural light. When making risk assessments, the value of early greenhouse studies can therefore be questioned.

OUT INTO THE FIELD

Today the focus is increasingly on field studies where net productivity is only one of the many processes that determine the growth of the plant. There are two basic designs in the experiments. One is to add extra UV-B with lamps that simulate different levels of ozone depletion, typically 10, 15 or 25 percent. This is the design in Abisko (Figure 4.3). At a couple of different areas in the birch forest, lamps are arranged to supplement the natural light. The times the lamps are turned on and off are adjusted to follow the natural change in UV-B radiation over the season. Very few of the designs take into account the different levels of light when clouds pass by. The reason is cost, but it is a potential problem, as it would lead to an overestimation of the loss in productivity.

The other basic design is to screen the natural radiation from the sun with plexiglass or with cuvettes containing ozone. In this set-up, it is obviously only possible to study the effects of UV-B levels lower that today's natural radiation. The ozone cuvettes also have limitations in the size of the plants and plant plots to be studied.

Most of the species that have been studied are annual crops. In about half of them, increased UV-B leads to a reduction in yield. For example, the yields of corn were reduced by 28 percent. Bean is another sensitive crop. Tomatoes seem to ripen earlier and wheat plants respond by not growing as high at ambient levels of UV-B compared to the plants grown under a UV-B screen. Other sensitive species are squash, mustard and black-eyed pea.

How do studies of loss in productivity translate into risk assessments? With difficulty, is one sure answer. Most of the negative effects are only seen when simulating rather severe ozone depletion, such as 20 percent. With a 10 percent ozone depletion, there might not be any effect at all on productivity.

It is also very clear that there are large species differences in UV sensitivity and that there is variation within the same species. For example, one cultivar of soybeans that has been grown under a simulated 25 percent ozone depletion during several consecutive seasons had its yield reduced by about 20 percent, whereas another cultivar actually increased its yield.

Figure 4.3   Lamps are used to simulate an increase in UV radiation at Abisko Scientific Research Station in northern Sweden. Photo: Carola Gehrke

BREEDING AND THE IMPORTANCE OF BIODIVERSITY

The difference in sensitivity between cultivars offers some hope to farmers, even if severe ozone depletion occurs. It shows that there is already genetic variability within the plants that should make it possible to breed for UV-resistant cultivars. A study of rice confirms the picture but is also a warning signal that not taking UV resistance into account in breeding programs might prove detrimental. Of 11 different rice cultivars that were grown under enhanced UV-B, about one third showed statistically significant decreases in total biomass and leaf area. Tiller number, which gives an indication of yield, was reduced in six cultivars. The most sensitive cultivars turned out to be two high-yielding varieties that were the core of the green revolution. Some of the most resistant cultivars were old landraces from areas with high natural UV-B levels, namely Tibet and Thailand. Protecting the genetic diversity within crop plants will thus be an important strategy in making it possible to breed for new UV-resistant strains.

OTHER ENVIRONMENTAL FACTORS

The effects of UV radiation on agricultural crops are also dependent on other environmental conditions. For example, it seems that the sensitive soybean is only affected when there is enough water available. In dry years, stress due to lack of water probably masks the effect of increased UV-B. This might be typical of how a plant responds to its environment. Its growth is an interplay between water, nutrients, temperature, carbon dioxide concentration and a variety of other factors. Each of them contributes to the productivity and also affects how sensitive the plant is to other factors.

Trying to predict plant productivity in a changing global environment, it might be especially important to learn how increased carbon dioxide concentration, climate change and increased UV-B will interact. Increased levels of carbon dioxide generally stimulate plant growth but in some plants the effect of carbon dioxide enrichment can apparently be cancelled by increases in UV. In one study of wheat, the fertilizing effect of carbon dioxide was clearly reduced. Soybean, on the other hand, was not affected at all. There might also be effects that are important for the competitive balance between different species. When grown under more carbon dioxide in combination with increased UV-B, a loblolly pine can change its growing patterns, with more dry matter in the roots compared to above-ground shoots.

At Abisko, a new experiment has recently been set up where the plants are subject to increased levels of carbon dioxide along with UV-B. It is too early to evaluate the effects of the treatment but it is one more step towards trying to get a grip on what will happen in nature when several environmental factors change at the same time.

FORESTS

In terms of total terrestrial biomass, forest trees are the most important plants to investigate. Unfortunately they are too big to study with many of the usual experimental designs and most experiments have therefore been carried out on seedlings. The results show that some species probably already suffer from UV stress at present radiation levels. In general, trees that are native to high elevations seem to be more tolerant.

The most important aspect of increased levels of UV-B on trees might be the cumulative effect over many years, as even a small decrease in biomass accumulation will add up to a substantial loss over time. Studies of loblolly pine have shown how there are decreases in the roots and in the shoots, even if there has not been a reduction in photosynthesis. The change in tree structure might also affect the competitive balance between different plants.

As in agriculture, it will be important to pay attention to species diversity. The situation now is that some trees have low natural diversity because of their evolutionary history and others are identical clones intentionally produced by the forest industry. The risk is that cultivated forests might end up consisting of one or two sensitive varieties that have to stand for ten to 40 years before they are harvested with a much lower yield than a tolerant variety.

THE ECOSYSTEM AS A WHOLE

Almost all studies on the effects of increased UV-B on plants have focused on individual species and only about 5 percent of all the studies so far have been done under natural field conditions. Even these have focused mostly on growth and gross morphological changes. However, the long-term consequences of a changing light environment might be much more subtle and work on the level of the ecosystem as a whole rather than on the individual plant (Figure 4.4). What happens, for example, when the buds of trees open earlier in the season? This has been shown to happen when oak, birch, beech and sycamore experience increased levels of UV-B. The undergrowth will definitely have a different light environment to deal with and it is not certain that the light and temperature conditions are still ideal for the same species as before.

There is also the interaction with insects and other animals. One example is the interplay between plants and pollinators. There is some evidence that increased UV-B can repress flowering. In petunia, the timing of the flowering is altered. If it no longer coincides with the activity of the pollinators it could be detrimental both for the reproduction of the plant and the survival of the particular pollinators.

Changes in the chemical composition of the plants have already been mentioned. This could change the behavior of predators and pests in a way

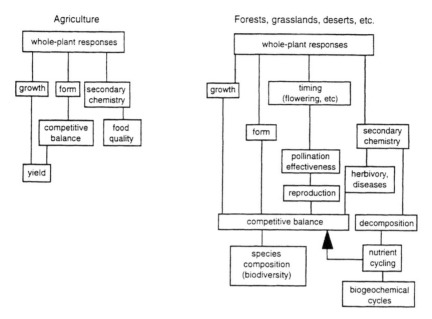

Figure 4.4   The direct effect of UV on plants has ramifications at many levels both in agricultural and natural ecosystems. Source: *Ambio*

that could, in turn, influence the predation or pest pressure on related species. Moreover, some predators might have to eat more of the plant material to get the same amount of nutrients, leading to a harder grazing pressure.

The most far-reaching consequences concern the interaction of the plants with the cycling of nutrients. Decomposition rates and nutrient content of the decomposing matter will influence how much of different nutrients become available for other plants. As nutrient levels in the soil often are limiting factors for plant growth, even small changes might have large effects.

Looking at ecological effects of UV damage to plants easily becomes a list of worst-case scenarios that are rather speculative in nature. It is no coincidence. There are hardly any studies of UV effects at that level. To get any idea of what the possible risks are, the natural step is to apply our understanding of known physiological and morphological changes in the plant in setting up investigations of whole ecosystems. It is in this context that the studies at Abisko become especially interesting. They add some substance to the speculations.

ABISKO REVISITED

At the end of 1994, four seasons have passed since the experimental plots were established on the birch forest floor. It will be several years before the plots will be finally evaluated to see if there have been any changes in species composition because some of the plants fare better than others under increased UV radiation. However, some results are clear. Deciduous shrubs seem to develop thinner leaves whereas the evergreens develop thicker leaves. The total growth rate is lower in the treated plots compared to the controls and the effect is most pronounced in the evergreens. They probably accumulate UV damage over time and it is clear that it takes more than a year before the effect can really be measured. Each of these plants has life spans comparable to trees and the real consequences for growth might take many years to evaluate.

The behavior of one of the moss species is still puzzling. A simulated 25 percent decrease in ozone leads to a 10–30 percent increase in growth of glittering feather moss (*Hylocomium splendens*). Ulf Johanson speculates that these kinds of results do not get much publicity because they are unexpected and do not fit in with an alarmist rhetoric. However, Carola Gehrke points out that the positive response may not last very long. It could also be a stress reaction and, if it is, the long-term effect might still be decreased growth.

Moreover, not all mosses react the same. A similar UV treatment of *Sphagnum*, which is the main peat-building moss genus, results in decreased growth. Such changes in growth could have significance for carbon storage in the biota of the ecosystem if they persist. Future summers will give an opportunity to measure the productivity in the moss mat as a whole. Each square decimeter consists of 400–600 individual plants.

Plants as a store of carbon and nutrients are also of interest in the experiments on decomposition. The release of carbon dioxide was 35 percent lower in leaves that had been irradiated with extra UV-B in the laboratory and in the first two months of decomposition there was a 5.6 percent lower relative mass loss. The microorganisms responsible for the decomposition are not only sensitive to the chemical composition of the leaves. Decomposing leaves that were left under UV-B lamps were not as well colonized with fungi. One of the common species did not colonize at all, a second colonized at a lower rate than normal, whereas a third species seemed unaffected. After a year, leaves grown under enhanced UV-B had a litter turnover that was 9 percent less than in the controls. After two years, the difference was 11 percent.

There is one process working in the opposite direction. UV-B directly promotes the decomposition of lignin, and this component of the plant mass was more easily broken down under enhanced UV-B. The total picture is that

UV-B slows down two processes of plant decomposition and speeds up a third process. The effects on the ecosystem as a whole will in the end depend on the magnitude of these changes.

WHO CARES?

Does it matter if there are changes in a natural ecosystem as opposed to a farmer's field or commercial forest, where we worry about yield? We certainly do not depend on the nature around Abisko for food, except for an occasional delicious cloudberry. In other areas, the economic value of the natural environment might be more or less significant, as is typical for ecosystems that are not managed but still play a role for informal economies. The role for supplying fodder might be more important. The subarctic forest certainly plays a key role for the semi-domesticated reindeer at the heart of the Saami economy. Regardless of present use, it is also very difficult to give an economic value to a particular natural ecosystem in comparison to the system that might evolve. The question is rather one of values and insurance against rapid change.

On a global level, one more concern enters the equation. Any change in the carbon-storing capacity of the system will be important for the atmospheric concentration of the greenhouse gas carbon dioxide and thus for climate change. If the system is disturbed, the storage capacity might decrease and reduce the effects of cutting emissions of carbon dioxide. Such effects might be the most difficult to predict, but if they occur, they will certainly have far-reaching consequences for human welfare. Some of these threads between plants, our atmosphere and climate will be further explored in the next chapter.

Chapter 5

Elusive threads in an intricate web

T HE DILEMMA WITH CHANGES IN THE GLOBAL environment is that the final outcome is not always easy to predict. Primary effects, such as damage to plants and plankton, can cause secondary effects. What happens, for example, if a key species is no longer able to fill its previous role in a specific ecosystem? What happens if a group of microbes responsible for recycling of nutrients becomes more or less active? Not only life is affected by ultraviolet radiation, but also the chemical characteristics of soil, air and water. It is an intricate web of inter-actions around one hot question: Will ozone depletion enhance climate change?

It is only in the past few years that researchers have seriously started to consider the large-scale effects of ozone depletion. This research is often gathered under the heading "biogeochemical cycles," which gives an indica-tion that everything is connected. Besides this self-evident claim, there are very few data on what could happen under different scenarios of future radiation levels. There is virtually no quantitative information available. However, a glimpse at some of the chemistry and biology involved should still give an appreciation of the importance of looking at nature across scientific disciplines. We can at least see the subjects involved in trying to tie the different problems of global change together. What do we need to ask before discussing whether ozone depletion will enhance or counteract the increasing greenhouse effect?

THE GREENHOUSE EFFECT

One of the driving forces behind the greenhouse effect is the presence of carbon dioxide in the atmosphere. Like other greenhouse gases, carbon dioxide traps the heat in the troposphere, rather than letting it reflect back into space. Without carbon dioxide and the other greenhouse gases the average temperature would be low enough to make Earth a frozen planet.

Currently the concentration of carbon dioxide is increasing at a rate that cannot be explained by natural fluctuations, which has caused concern about global warming. Based on sophisticated computer models scientists have predicted a global temperature increase of somewhere between 1 and 3.5°C by the year 2100 compared to pre-industrial times. A few degrees might not seem like much, but it translates into more far-reaching climate changes than we have experienced since the last ice age. At a regional level, the climate changes might mean different patterns in rainfall with risks for droughts or floods, more severe storms, and a shift in temperature with concurrent effects on plant growth, productivity and nutrient cycling. The first signs are already gathering with increased average temperaratures, slight shifts in seasons and disrupted weather patterns.

The major cause of the increasing carbon dioxide levels is emissions from the burning of fossil fuels. Clearing of forests also carries a share of the blame. Reduction of forests means that less carbon is bound in the biomass of trees and undergrowth, leaving it in the form of carbon dioxide in the atmosphere.

THE CARBON CYCLE CONNECTION

The concentration of carbon dioxide in the atmosphere can also be described as a function of the activity in different parts of the carbon cycle. All plants and photosynthetically active plankton harvest carbon dioxide and bind it as organic carbon, such as sugar, carbohydrates and cellulose, in the photosynthetic process. It is returned to the atmosphere when the plant itself, animals or microorganisms use the energy-rich carbon compounds as food. The carbon can also be returned if the organic matter breaks down in chemical processes, such as burning or decomposition by light (Figure 5.1).

The amount of carbon that is bound and released in this natural flux is much larger than the anthropogenic emissions. Each year, terrestrial biota accounts for about 100 gigatonnes (Gt) of carbon dioxide exchange with the atmosphere, which is an order of magnitude higher than the carbon dioxide that can be traced to the burning of fossil fuels. The oceans contribute another 100 Gt to the carbon flux. Measurements of carbon dioxide concentration from Mauna Loa, Hawaii, make the role of plant life very visible. The variation corresponds to the alternations of the seasons and shows how the biosphere accumulates carbon over the summer and releases it during the winter months (Figure 5.2). Fossil fuels, unfortunately, only release carbon.

The carbon flux from the biosphere is generally in balance—over the year uptake equals release. However, the size of the flux makes even small changes matter. If plant life should decrease its overall carbon-storing capacity, each small increment could have a large effect on the carbon dioxide concentration in the atmosphere.

This is where ozone depletion and the effect of UV-B enter the scene. There is no question that increasing levels of UV-B can decrease the carbon-catching capacity of individual plants. The key question in relation to the greenhouse effect is whether increased UV-B will also affect the total amount of carbon bound in biota. As was described in the previous chapter, science has yet to come up with the quantitative data to answer this critical question. One of the few statements that have been made comes from the scientific team on the

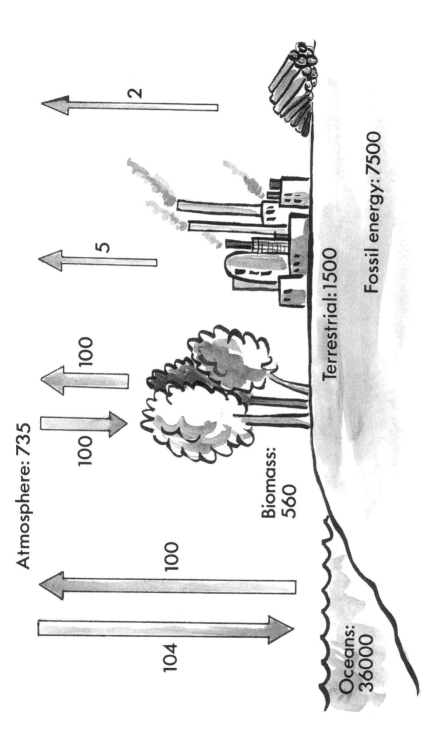

Figure 5.1 Both human and natural sources contribute to the cycling of carbon. The numbers indicate the amount of carbon from each source (in Gt). The net increase in carbon dioxide in the atmosphere comes mainly from the burning of fossil fuels. Illustration adapted from *Ambio* by Lena Adamsson

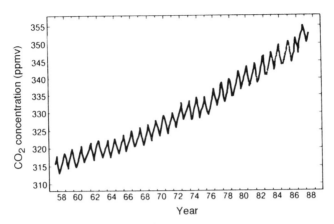

Figure 5.2   Carbon dioxide levels in the atmosphere oscillate with each season. The trend, however, is a steady increase as shown by these measuremnts from Mauna Loa, Hawaii. Source: *Climate Change. The IPCC Scientific Assessment*

Icecolors cruise during the ozone hole of 1990. They measured a decrease in primary productivity of the phytoplankton of 6–12 percent, but concluded that this would not be enough to perturb the global carbon cycle.

The most important aspect might not be primary productivity. Most of the plankton and plants are eaten or decomposed and the carbon returned to the atmosphere within a short time. To affect the carbon cycle, the standing crops have to increase. It could be compared to more trees in the forests of the world. The other way changes in primary productivity could affect the levels of carbon dioxide in the atmosphere is if organic matter is removed permanently from the biosphere. One example of this could be dead plankton that settle on the ocean floor without being consumed. In Antarctica this is a very small proportion of the plankton mass. Another example could be carbon bound in *Sphagnum* moss that end up as peat in the permafrost, where there is no biological activity to decompose the stored plant matter.

SENSITIVE STORES

The focus on carbon storage points to some other mechanisms that might be important for the interaction between ozone depletion and the greenhouse effect. Aquatic environments all contain organic matter in forms that are not easily available for animals or microbes. Humic acids, which color the water brown, are one example. Some of these substances can be broken down by UV-B. Once the carbon-rich humic acids are bite-size pieces of food, microbial life in the water would easily give them a passageway into the atmosphere by their metabolism and respiration. A terrestrial equivalent is the UV degradation of lignin in plant litter, which makes the stored carbon more available to fungi and other decomposers.

The effect of UV-B does not stop at this level. Let us follow the thread provided by the focus on humic acids one step further. As humic acids decompose, the water should turn less yellow and in clear water UV-B will penetrate deeper. It could thus affect carbon-harvesting phytoplankton that otherwise would be protected. Penetration of UV-B to greater depths would also strike bacterioplankton, which are responsible for decomposition. They add carbon dioxide to the system and this addition would be potentially hampered. It has been shown that both freshwater and marine bacteria are sensitive to UV-B and that the rays from the sun reduce the growth of bacterioplankton in the upper layers of the ocean already at present UV levels. It is not straightforward to estimate the total effect of UV-B penetrating deeper into aquatic environments but the net effect would depend on the relative sensitivity of producers and users of carbon dioxide.

In terrestrial environments there are several important carbon stores that can be affected by changes in UV-B. The first examples are the living plants and changes in competitive balance between different plants. If deciduous species are favored over evergreens, less carbon would be bound in biomass during the winter. Experiments with dwarf shrub vegetation at Abisko in northern Sweden show that such changes are rather likely.

Other long-term carbon stores are leaf litter and peat. Any environmental change that makes this carbon more available to microbes or other decomposers would increase the carbon dioxide concentration in the atmosphere. One such example is if UV-B has a negative effect on the growth of mosses. As the ground in cold areas would no longer be as insulated, the increased ground temperature during the summer would stimulate microbial activity.

NEW APPROACHES

Do the effects on carbon cycling on an individual species or at a community level have any significance for the global climate? The question is impossible to answer with present knowledge. What is even more problematic is that very little research aims at trying to answer this question. Learning more about what goes on in a plant will not be of much help. Accurate measurements of primary productivity or decomposition within a whole ecosystem will be important, but only if there is also information about the amount of carbon that is removed or added to the active cycling. Some researchers point to the need for developing satellite imaging techniques as a way to make large-scale pictures of the productivity. Such technology is already used to measure chlorophyll as an indication of phytoplankton growth in coastal areas. A complement could be tethered or free-floating buoys that constantly measure the incoming radiation along with the color of the water as an indication of plankton productivity. The information is conveyed via satellite. The next step would be to integrate this kind of data into the computer-based Earth system models that atmospheric scientists use to predict the future climate.

NUTRIENTS THAT LIMIT AND PROMOTE GROWTH

In the fall of 1993 a spectacular experiment was performed outside the Galapagos Islands. Tonnes of iron were dumped into the ocean to tease plankton into increasing their photosynthetic activity. It is known that iron is a limiting factor for plant growth in some areas of the ocean and the experiment had its origin in the controversial idea of mitigating global warming by fertilizing the plankton. The plankton did indeed respond to the treatment and the plankton mass increased. The first year, the effect was temporary. Zooplankton and other creatures appreciated the human effort to stock their cabinets and quickly munched all the extra food. The net effect on carbon storage in the surrounding sea turned out to be very small. A second trial in 1995 was more successful in actually getting a sustained growth of phytoplankton.

Before the experiment was launched, criticism was very loud because of the risks involved in messing with the ocean ecosystem. The researchers themselves are concerned about how eco-engineers will eventually use the results from the successful 1995 experiment. However, regardless of political criticism, the experiments have interesting scientific implications. They show how important trace elements can be for the productivity of a marine ecosystem. Other trace elements can play similar roles. Manganese is one such limiting factor in many marine systems.

What would happen if some environmental factor increased the availability of iron and manganese on a global scale? This factor could be increased UV-B. Solar UV radiation has the potential to change the oxidation state of iron and manganese if there is oxygen present in the water. This could help make insoluble oxides and hydroxides of these metals more available to the phytoplankton. Moreover, UV-B inhibits microbes that convert the soluble forms of manganese to insoluble oxides.

One of the key nutrients in many ecosystems is nitrogen. For example, the productivity of large parts of the northern boreal forest is limited by nitrogen deficiency in soil. Any change in this nutritional status could potentially change the productivity of the forest. The typical effect of increased UV-B would be if there are changes in the rate of decomposition of the litter on the forest floor. These kinds of changes have been suggested in the work on decomposition of subarctic brush vegetation in Abisko.

SULFUR FOR THE ATMOSPHERE

Carbon dioxide is not the only gas that will determine the magnitude of the greenhouse effect. A look at the biogeochemical cycle of sulfur offers another glimpse into the complexity of the interaction between ozone depletion and global warming.

The sulfur compound dimethyl sulfide (DMS) plays a key role in the formation of thin clouds over the ocean by providing a condensation nucleus for the water vapor in the atmosphere. The main source of DMS is a group of phytoplankton called coccolithophorids. In remote marine areas they account for as much as 90 percent of the DMS production and on a global scale for about 15 percent.

It is well known that UV-B affects phytoplankton in general, but what does it do to the coccolithophorids? If they are more, or less, sensitive than other species, it could cause a change in the production of DMS and thus a change in these stratiform clouds. These clouds help cool our planet and a change in their distribution or optical properties could thus affect the climate.

Another sulfur gas produced in the sea is carbonyl sulfide. This gas is a major source of sulfate particles in the stratosphere, which in turn play a role for ozone depletion. The sulfate particles act in the same way as ice particles and provide a surface for the catalytic reactions that can speed up ozone depletion. In some areas of the ocean, such as the North Sea and the Gulf of Mexico, the production of carbonyl sulfide seems to be sensitive to radiation in the UV-B range. If it holds true on a large scale, ozone depletion could potentially provide more of the catalysts that speed up the destruction of stratospheric ozone—a positive feedback loop.

NITROGEN GASES

Another gas from the terrestrial ecosystem that has received some attention in the discussion of biogeochemical cycles is nitrous oxide (laughing gas). Nitrous oxide is a potent greenhouse gas and increased production would add to global warming. Moreover, nitrous oxide is involved in some of the chemical reactions that lead to destruction of stratospheric ozone. How would nitrous oxide be affected by increased UV-B? One route could be through changes in the nitrogen-fixing and nitrogen-releasing activities of microorganisms in the ground. Research on subarctic dwarf shrubs has shown that UV-B-exposed litter has higher levels of another form of nitrogen, namely ammonia. It is an indication that the microbial activity can change, even if it is not enough for making any prediction about changes in nitrous oxides on a large scale.

ELUSIVE THREADS IN AN INTRICATE WEB

This chapter has probably not been easy reading, in spite of the simplified descriptions. It is no coincidence. The threads in the intricate web of biogeochemical cycles demand an understanding of both biology and atmospheric chemistry in such detail that very few people have even a chance of getting a complete picture. The lack of answers to simple questions makes it even messier. The take-home message is that it is important to study the interactions between different systems and different environmental problems. However, it is difficult to prove any claim that increased levels of UV-B will worsen global warming. There are simply too many actors involved moving in too many different directions. The next chapter will introduce even more connections between ozone depletion and climate as we step into the realm of atmospheric chemistry.

Chapter 6

A breath of fresh air?

A HOT SUMMER DAY IN THE CITY. SINCE EARLY morning the cars have spewed exhaust fumes and by the beginning of the afternoon the air feels thick to breathe. A visitor arriving by air can see the lid of dirt—a yellow haze that covers the entire area. The picture as well as the smell are all too familiar. It is smog, caused by the photochemistry of polluted air. And the pollution stretches far. In fact, rural areas around large cities can be even harder hit than the city centers. It can be difficult to get a breath of truly fresh air anywhere within the large industrial regions.

One of the main components of photochemical smog is ozone. It is the same gas that is depleted in the stratosphere, but at ground level ozone is increasing over many populated regions. Ozone is a noxious gas and in some places the levels are high enough to regularly exceed the standards for acceptable air quality. And people suffer. The direct effect is an increased rate of respiratory disease and even deaths because of asthma attacks. Plant life suffers as well.

There are several routes by which stratospheric ozone depletion can worsen the air quality. One of the most important is the direct effect of UV-B on the chemistry of the troposphere. The role of UV-B goes back to its ability to break the chemical bonds of gases that are present in the troposphere such as nitrogen oxides, nitric acid, hydrogen peroxide, formaldehyde and ozone. The result of this chemical interaction is that there will be an increasing number of very reactive chemicals in the air, so-called radicals. They, in turn, drive other chemical reactions.

How this photochemistry affects the air quality will depend on what other pollutants are around. At certain concentrations of nitrogen oxides and volatile organic carbon compounds from different combustion processes, there is a production of ozone. With more UV-B, the levels of pollutants necessary to drive this reaction do not have to be as high as before. It means that photochemical smog will form earlier in the day in many cities. This kind of pollution will also spread over larger areas. To maintain air quality, there will have to be more stringent controls of emission of volatile compounds in a situation where many cities have problems meeting even the present regulations.

GLOBAL TROPOSPHERIC OZONE

Looking at the production of ozone in the troposphere as a whole, the conclusions are not as clear as in polluted areas. Here the problem is not mainly one of health but of the role of ozone as a greenhouse gas. Increased levels would add to the greenhouse effect. UV-B is active both in the creation and destruction of ozone and it has been difficult to calculate the predominant direction with certainty. Different computer models of atmospheric chemistry give different results. As long as the levels of nitrogen oxides are low, there is mostly a decrease in ozone in the models. In Antarctica, the levels of tropospheric ozone did indeed decrease during the spring and summer seasons of 1976–89, which was probably caused by the increased UV-B levels under the ozone hole.

When nitrogen oxide levels are higher, as in most industrial regions, ozone concentrations tend to increase. The difficulty is in predicting the nitrogen oxide levels and other factors that steer the chemical reactions in the different directions.

Looking at the globe as a whole, the concentration of tropospheric ozone varies a great deal both regionally and vertically. It has made it very difficult to estimate what the global trends have been so far. The most pronounced increase in tropospheric ozone has been over northern industrialized regions, which can be expected considering the amounts of nitrogen oxide pollution.

CLEANING ABILITY AND OTHER GREENHOUSE GASES

In relationship to global warming, there is one positive effect of an increased reactivity of the troposphere. It relates to the greenhouse gas methane. Methane is removed by one of the radicals that is expected to increase with higher UV-B levels. In fact, the previous increase of methane in the atmosphere has slowed down and a substantial portion of this trend can be attributed to increased UV-B levels. A more reactive atmosphere also helps clean out some of the CFC substitutes, namely those with an added hydrogen.

The chemical reactivity of the troposphere also has implications for environmental problems via the biogeochemical cycle of sulphur. Most important is that there will be more sulphate particles, which play a role in cooling the Earth. The particles act as parasols. Over polluted areas in the northern hemisphere, they have masked much of the effect of increased levels of greenhouse-enhancing gases.

TRACE GASES FROM PLANTS

Many of the trace gases that are important for the chemical reactions in the troposphere have a human origin. However, some of them can also come from nature. For example, plants produce carbon monoxide and non-methane hydrocarbons.

It is difficult to tell how much UV-B could modify the concentration of the biological production of these chemically active trace gases. However, plants as a source are important enough that any disturbance of the plant community could alter the concentrations in the atmosphere. One route would be through changes in species composition of a plant community. Moreover, dead leaves seem to produce carbon monoxide much faster under high UV-B levels.

Plants also act as scavengers of trace gases. Humid tropical forests can take up significant quantities of nitrogen oxides as well as carbonyl sulfide.

STRATOSPHERIC OZONE

Will deletion of the ozone layer in itself contribute to the greenhouse effect? Stratospheric ozone is indeed not only interesting because of its effects on UV-B radiation. Ozone is also a greenhouse gas with an ability to trap heat, and less ozone in the stratosphere will counteract the greenhouse effect. It has even partially offset the direct greenhouse effect of CFCs and halons, which in themselves are quite potent greenhouse gases.

Again, it should be clear that there are forces driving in opposite directions in discussing ozone-depleting substances and ultraviolet radiation. And just as in the previous chapter, it is difficult to draw any clear conclusions about whether the sum of all the different forces will worsen the greenhouse effect or mask it. What is clear, however, is that the increase in UV-B that follows ozone depletion will worsen air quality in many industrialized areas around the world. The price is an increased number of deaths from asthma, chronic respiratory disease and direct ozone damage to plants.

Chapter 7

Red alert

S HE WAS 92 YEARS OLD WHEN SHE FELT THE ITCH ON her ear, just by the hearing aid. At first she though it was a chafe, but the small lesion in the skin did not want to heal. A consultation with the doctor confirmed her fears. It was a small skin tumour. It was the second time. Five years earlier she had had a spot on her nose removed for the same reason. The doctor had said it was routine surgery and nothing to be worried about, but she had developed a large infected sore in the middle of her face and it took a long time before the whole ordeal was over. This time, surgery would mean removing the ear, she was told. Instead, she had radiation treatment. Talking about her skin cancers, she remembers a summer in her youth. It was unusually sunny and warm. Her hat had disappeared and she suffered some bad sunburns.

This is not just an anecdote of the trials and tribulations of growing old. If ozone depletion continues and ultraviolet radiation increases, the story of this old woman will strike close to home for many more people, even if they do not live to be 90. The anecdote is really about how exposure to the UV rays from the sun increases the risk of contracting skin cancer. For the non-melanoma skin cancers, the evidence is compelling and there are estimates that each percentage decrease in stratospheric ozone will lead to a 2 percent increase in the incidence of these cancers. If ozone depletion reaches 10 percent over a sustained period of time, an additional 250 000 people would be affected each year. Even if international agreements about CFC phase-out ensure that the ozone layer returns to normal, the incidence of the non-melanoma skin cancers could be 25 percent higher in the year 2050 than in 1980 (Figure 7.1).

The non-melanoma skin cancers are usually treatable and it is rare that people die from them. Nevertheless, each affected person will have to deal with surgery or radiation treatment to keep the cancer from becoming a major health problem and each case will be an additional burden on the health care system. Moreover, some groups might be much harder hit. These include people with a hereditary susceptibility to cancer and anyone with a compromised immune system, such as people infected with the HIV virus and those who have had organ transplantations. Light-skinned people living in areas that lack well-developed medical care are also at extra risk.

MAKING THE CONNECTION

There are two ways to understand the connection between skin cancer and exposure to UV radiation. One is studying large groups of people to look at their past exposure to the sun and relating it to the probability of them

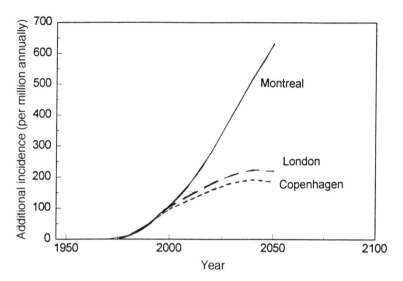

Figure 7.1   The incidence of non-melanoma skin cancer in the future will depend on the success in phasing out ozone-destroying chemicals. The graphs are based on the Montreal Protocol, its London and Copenhagen amendments and on statistics from the Netherlands using a model developed by the Dutch National Institute of Public Health and Environmental Protection. Source: *UV Radiation from Sunlight*

contracting cancer. The second route is to gain a basic knowledge of what happens in the skin—what factors are involved in triggering the cancer. Looking at the different events when the rays from the sun hit the skin will also help us understand why it is not enough to avoid a sunburn. In fact, the mechanisms for sunburn and cancer turn out to be very different.

RED AND TENDER

Back in the 1950s, a young Dutch scientist named Jan van der Leun started to illuminate skin with different light sources to see what happened. The most obvious effect was the reddening that appeared after a couple of hours. Most people recognize this as sunburn and know from experience how painful it can be. Jan van der Leun was curious about the mechanism. The scientifically intriguing puzzle was that nothing was visible at first. It was only after some hours that the precise area that had been illuminated turned red. What was going on? What is a sunburn?

Today this Dutch scientist is the chairman of the scientific panel responsible for assessing the environmental effects of ozone depletion under the Montreal Protocol. He still does not have a satisfactory answer to the question he posed in his youth, however. The current understanding is that light hitting the outer layer of the skin, the epidermis, triggers the production of some substances which diffuse into the dermis below. The dermis is filled with blood vessels, and the chemical substances cause them to dilate, making the skin red and warm to the touch. The time it takes for the chemicals to diffuse would explain the delayed reaction (Figure 7.2).

Do the same reactions cause the pain after a sunburn? Not necessarily. The pain is probably a reaction to UV radiation hitting nerve endings close to the skin surface, which makes them more sensitive to heat. The peeling of the skin that comes after a sunburn could be yet another unrelated event. These are probably cells that have been so severely damaged that they die.

Will people get more sunburns if ozone is depleted? The answer is not evident. It is true that UV-B is much more effective than longer wavelengths in causing sunburn and this is the part of the spectrum in which the increase in radiation will take place. One of the standard action spectra used in photobiology is the erythema spectrum showing neatly how the reddening of the skin is related to different wavelengths. I have seen the effect first hand on some people on board the research vessel *Nathaniel B. Palmer*. We were at Anvers Island by the Antarctic Peninsula visiting an American research station. The weather was clear—a beautiful spring day with snow and water reflecting the intense sunlight. It also happened to be a day with very low ozone. Some of the people on board, who spent a little too much time outside, suffered severe sunburns. I could feel the effect myself in spite of sunscreen and spending only a few hours outside.

Some UV meters, namely the Robertson–Berger meters, weight the radiation according to the erythema action spectrum and these readings thus give a good measure of the sunburning capacity of the radiation on a particular day. Most UV indexes that are used to warn people about the damaging effects of the sun are based on the erythema spectrum.

In spite of these connections, ozone depletion might not really be a problem when it comes to sunburn according to Jan van der Leun. Emigrant studies have shown that the skin can adapt to higher UV-B exposures in new environments. This suggests that the same will happen in, for example, northwestern Europe when the UV-B irradiance gradually increases.

More important, however, is how people up north deal with the sun during the transition from winter to summer, when the minimal erythema dose increases 10 to 20 times. Most people handle this by trying to adapt the skin by a series of exposures, getting only a little sun at a time. A complete adaptation

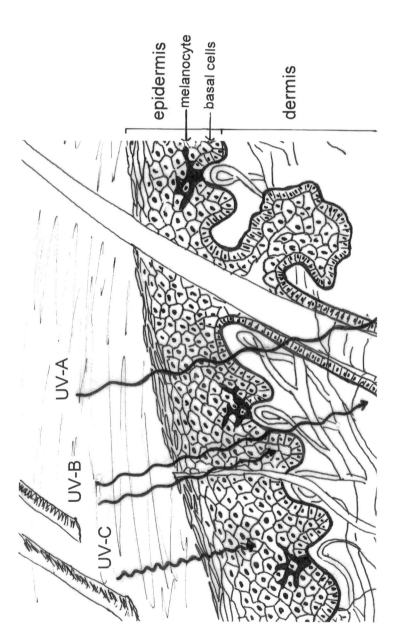

Figure 7.2 The penetration of UV radiation into sensitive cells in the skin depends greatly on the wavelength. At a depth of 70 micrometers, corresponding to the location of the basal membrane, 19 percent of UV-A (365 nm), 9.5 percent of UV-B (313 nm) and less than 1 percent of UV-C (254 nm) remains

of the skin to the summer sun requires about 15 exposures. If the UV irradiance increases by 20–30 percent because of ozone depletion, it would still only require approximately one more exposure for the skin to adapt, according to Jan van der Leun's reasoning. Regardless of ozone depletion, the trick to avoiding sunburn is to go through the spring adaptation slowly.

Do sunburns cause cancer? Public education campaigns to prevent skin cancer often warn people about getting burned by the sun. The anecdote in the beginning of this chapter also points to a link between the painful but temporary red skin and the long-term effect in the form of an increased risk for cancer. However, the connection between the two events is not straightforward. First of all, there are different kinds of skin cancer, and not all of them have a clear connection to sunburn. Secondly, it is not necessarily the sunburn itself that is the cause of cancer, even if getting a lot of sun is involved in both events. Understanding the difference is important for anyone trying to prevent skin cancer.

SKIN CANCER

Cancer can be described as a disease where the body has lost control over how certain cells divide. Normally, there are mechanisms within each cell that regulate the growth, division and even death of the cell. If these systems fail, and the out-of-control cells are not caught by the immune system, a cancer can develop. Skin cancer can involve many different cell types in the skin, such as basal cells, squamous cells and the pigment-producing melanocytes.

The most common and best studied forms of skin cancer are the non-melanoma cancers. This is really a group of different diseases where the most common are cancers of the basal cells and of the squamous cells in the epidermis of the skin. Both cancers are typical of old age and can usually be treated successfully. Less than 1 percent of people who fall ill actually die from non-melanoma skin cancer, most of them from squamous cell carcinoma.

DAMAGE TO THE BLUEPRINT OF LIFE

Skin cancers are initially caused by damage to the DNA in the skin cells. The DNA is the hereditary material and it contains all the information the cell needs to function, including the blueprint for many signals that control growth, division and death of the cell. The DNA is very sensitive to UV radiation, specifically in the UV-B range. For example, the radiation can cause the molecule to lock itself up and the information next to these "knots" may become scrambled or simply deleted. If it happens to messages that are involved in controlling how the cell grows and divides, such damage might trigger a cell into becoming a cancer cell.

Ultraviolet damage to DNA is quite common and the damage occurs at much lower doses of UV radiation than is necessary for a sunburn. In real life, every exposure of skin to sunlight causes thousands of DNA defects. Fortunately, very few of these cells get any opportunity to develop into a cancer. The cell can usually repair the damage before the cells divide and any harm is done. Sometimes the repair system cannot keep up with the damage, however, or has trouble repairing certain types of injuries. This leads to a slow accumulation of defects in the DNA. In fact, a clue to the mechanisms behind skin cancer originally came from patients with a hereditary defect in their DNA repair. They are much more likely than other people to suffer from all kinds of skin cancer because their DNA damage is left unmended.

How would UV-B increase the risk of a cell turning cancerous? Part of the explanation is that the amount of DNA damage increases, making it more difficult for the cell to repair the damage before it divides. Over time damage can accumulate, increasing the likelihood that it will also affect genes that control growth, division and death.

The most important molecular evidence for the role of UV-B in skin tumors comes from signs of damage in one of the genes that control cell division. The gene is called p53 and normally stops the cell from dividing to leave time for the repair system to look after the damage. More than half of all squamous cell and basal cell carcinomas have some damage, a so-called mutation, in this gene and the specific damage is almost a signature of UV-B having wreaked havoc with the hereditary material. Studies on mice show that the p53 gene is altered already in the pre-cancerous lesions of squamous cell carcinomas, which implies that it is one of the early events in the development of a tumor.

In the human body, the p53 mutations seem to accumulate over time as more and more cells have been exposed to the sun. Looking at sun-exposed skin, it is possible to actually measure an increase in p53 mutations compared to skin on the buttocks, which are very seldom exposed. With enough mutated cells, there is an increased risk that one of them will eventually go haywire.

The body has several lines of defense to avoid damage by mutated cells. One is that the cell can self-destruct in a programmed cell death. Recently, p53 has also been implicated in this process. Maybe this is a last line of defense for the body; when the cell is unable to repair the DNA damage, it is better to self-destruct than to turn into a cancer. If the theory holds up, we should be thankful for the skin peeling off.

The most important defense, however, is that the immune system has effective mechanisms for recognizing cells that do not behave normally and for removing them. As will be described in chapter 10, UV-B can unfortunately hamper the immune system, making it more likely that a pre-cancerous cell will escape its attention. The effects on the immune system are also apparent at UV doses well below what is needed to get sunburned. This means that two of the major factors involved in non-melanoma skin cancer, the initial DNA damage and the hampering of the immune system, occur at UV doses well below those that cause a sunburn.

OF MICE AND PEOPLE

The picture that lifetime exposure to the sun is a risk factor for non-melanoma skin cancers, rather than sunburns alone, fits very well with what emerges from epidemiological studies. It is very clear that people who have worked outdoors during their whole lives, such as fishermen and farmers, are more likely to contract non-melanoma cancer, than indoor workers. Moreover, people living in sunny areas are more likely to get sick than those from latitudes with less intense sunlight.

Animal studies also confirm this picture. Most animals can get some of the non-melanoma skin cancers and mice are often used as a model for what would happen to people. The difference from the epidemiological studies is that the animal studies make it possible to control the doses of UV and the exposure times. Again, it is the total lifetime dose of UV that increases the risk for non-melanoma skin cancer and the animals do get cancer at exposures that are well below what is needed to get a sunburn.

Some of the human population studies have looked at people who have migrated to sunny areas early or late in life. There are some indications that getting a lot of sun at a young age might increase the risk more than getting the same amount of sun at an older age. This is quite logical considering that the cells in a growing child divide much faster than in an adult, increasing the risk that damaged DNA gets copied and carried on into new cells. From a prevention point of view, it makes it especially important to protect children from the sun.

HOW BIG A RISK?

Looking at the mechanisms of skin cancer does not say anything about the magnitude of the risk and how the depletion of the ozone layer would contribute to the incidence of cancer. Epidemiological data can be of some help. For example, population-based studies of lighter-skinned people in the northern and southern United States have been used to calculate the increased risk of getting more sun. The exposure is assumed to follow latitude. But there are problems with this approach. The assumption is that factors other than exposure to sunlight are equal for the different groups of people and it is difficult to know whether this is true. Moreover, people's behavior also affects how much sun they get. The studies assume that a population's behavior is the same regardless of where we live.

Another way to estimate the increased risk from ozone depletion is to gather information from experimental animals. The animal studies make it possible to look at what wavelengths of UV radiation are the most damaging. This can be used to calculate the increase in carcinogenic UV dose caused by ozone depletion. The approach gives a safer quantitative base for risk assessment, but of course assumes that the carcinogenic process is similar in mice and humans.

The relationship between damage and the wavelengths is called an action spectrum (Figure 7.3). The action spectrum for skin cancer differs from the action spectrum for DNA damage because the radiation is moderated by the

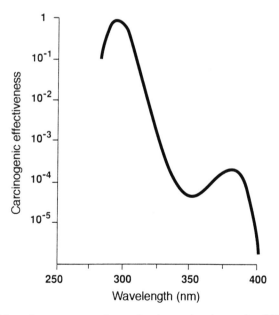

Figure 7.3   This action spectrum determined on mice shows that UV-A plays a role in the induction of squamous cell carcinoma. Source: *Effects of Increased Ultraviolet Radiation on Biological Systems*

skin itself before it reaches the DNA. This also causes differences between human and mice action spectra as people have thicker skin. Action spectra also incorporate all the other events involved in moderating the onset of cancer, such as DNA repair, cell death and immune responses.

UV-A AS A CO-CULPRIT

Working out the action spectra for skin cancer has raised some new questions about the etiology of skin cancer. One of the most important findings from the laboratory studies is that UV-A also contributes to the carcinogenic process. The mechanism for this is not known as DNA itself is hardly affected at all by UV-A. However, it means that the increased risk for contracting skin cancer because of ozone depletion has been adjusted downward. The reasoning is that ozone only affects the amount of UV-B, which has become relatively less important as a risk factor.

It should be noted that recognizing the role of UV-A does not change the risks associated with being exposed to the sun. Sun that reaches Earth's surface contains much more UV-A than UV-B and even if the carcinogenic effect of UV-A is only one ten-thousandth of UV-B, 15–20 percent of the carcinogenic effect of sunlight comes from UV-A. The take-home message is that protection from the sun is a good idea regardless of ozone depletion. Moreover, sunscreens should block both UV-A and UV-B to give good protection against skin cancer. It also means that solariums are not safe just because they have only UV-A in their lamps. They can actually be more dangerous, as is discussed in the section "A healthy tan?".

Back to risk assessment for ozone depletion. Based on the action spectrum for human skin and calculations of how ozone depletion will affect UV radiation, the current models indicate that each percent decrease in ozone will lead to a 2 percent increase in the incidence of non-melanoma skin cancers.

Referring to the northern hemisphere, the risk scenario for skin cancer has been compared to having the population move 400 kilometers south or 500 meters up in altitude. This would be true if one looks at the increase in risk for a single individual. Looking at populations is quite different.

"The Dutch population could move to an altitude of 500 m only by going up in the air, and 400 km south only by invading Belgium and France. How can we know that the risk of such moves would be small," Jan van der Leun comments.

Looking at what happens in real life is sobering. Many of the data on risks derive from migration studies of people who have moved to Australia and

Israel, where skin cancer is a major public health concern for the white population. Indeed, the migration studies have led to the conclusion that a 20 percent increase in the incidence of non-melanoma skin cancers can be expected after several decades as a consequence of ozone depletion.

CHANGING NUMBERS

Looking at the United Nations Environment Programme's risk assessment for skin cancer one can see how it has moved steadily downward from the estimates of about 1980, when the first numbers were presented. At that point in time, it was calculated that each percent decrease in ozone would give rise to a 4 percent increase in skin cancer. This, in turn, was lower than a number that circulated in the beginning of the 1970s, 6 percent. Others had, however, presented more conservative estimates based on essentially the same data.

Are these changes a sign that science cannot be trusted? I would say no. However, it shows that risk assessments are only temporary and only as good as the knowledge on which they are based. The first estimates were made by individuals and based on rather simple atmospheric models and knowledge about non-melanoma skin cancer incidence in different parts of the United States. The risk was refined downward in the late 1970s when more careful calculation gave a better estimate about the relationship between ozone depletion and increase in UV radiation. Moreover, this was when the first epidemiological studies were published, making it possible to look at the relative risk depending on how much sun people were exposed to.

By the early 1990s, the effect of UV-A had turned up in the animal studies and was entered into the calculation making the risk associated with ozone depletion seem even lower than before. Each percent decrease in ozone would give a 2.3 percent increase in the incidence of non-melanoma skin cancer, according to the 1991 UNEP report. In the most recent assessment, which came in 1995, the mice data have been adjusted to account for the differences between human skin and mice skin and the number was reduced to 2.0. In other words, ozone depletion will still cause an increase in skin cancer but the increase is likely to be less steep than was thought 25 years ago.

Will the numbers keep on adjusting downward? Jan van der Leun, who is one of the scientists responsible for calculating the numbers, says that he is intuitively confident that they are now converging on the real value. The joker in the calculations would be if there are some yet unknown mechanisms that have not been taken into account. But it would have to be something that has not been intrinsically taken into the epidemiological studies or in the animal experiments. The discovery of the role of the immune system for carcinogenesis did not change the risk assessment, whereas UV-A entering the equation played a role because ozone depletion almost only affects UV-B.

MELANOMA SKIN CANCERS

The risk assessments for skin cancer in connection with ozone depletion have only concerned the non-melanoma cancers, so far. A much more serious form of skin cancer involves the pigment-producing cells of the skin epidermis, namely the melanoma skin cancers. These cancers often affect young people, they spread quickly and can invade other parts of the body if they are not detected early and removed. The melanoma skin cancers have increased rapidly in all white populations. In Scandinavia the incidence of malignant melanoma is currently increasing by 6 percent per year. In many countries the incidence has doubled or tripled during the past 30 years and, in spite of better treatment, more people die from malignant melanoma today than 30 years ago. In parts of Australia, malignant melanoma is the fourth most common cause of premature death. The problem is clearly very serious and numerous public education campaigns have been started to try to teach people how to identify pre-cancerous spots on the skin and how to behave in the sun to decrease the risks (Figure 7.4).

The connection between exposure to the sun and melanoma is not as neat as for the non-melanoma skin cancers. Outdoor workers do not seem to be hit as hard as indoor workers and the cancers are likely to appear on the trunk and legs rather than on the face, neck and arms, which are most often exposed to the sun. Nonetheless, the risk for melanoma is greater in sunny areas than in areas with less sun, when looking at people with the same type of skin.

The major clue to the cause of malignant melanoma has come from studies of people who have immigrated to Australia either at a younger or at an older age. The risk for the melanoma skin cancers is much higher in people who came when they were young, which has spurred the advice to keep children protected from the sun. Surveys of people with malignant melanoma also indicate that they are more likely to have had severe sunburns as children. The studies have many weak points, such as selective memory of people filling in the questionnaires. Nevertheless, sunburn might play a role, especially at a young age. The role of occasional high exposures is also supported by the fact that malignant melanoma is more common in higher socio-economic groups. They have been able to afford going to sunny areas on their vacations to burn parts of their bodies that are not used to the increased solar range.

Figure 7.4 "Tan slowly." The Swedish Cancer Foundation is working to raise people's awareness about the dangers of being careless with the sun. During the summer of 1995, they visited beaches to give people advice and a chance to have medical professionals check the skin for potential malignant melanomas. Illustration: Nils Peterson, courtesy of the Swedish Cancer Foundation

Will ozone depletion increase the risk for melanoma skin cancers? This question is very difficult to answer. As there are differences in the patterns of who is at risk for non-melanoma and melanoma skin cancers, it would be risky to draw any conclusions from previous risk assessments and apply them to the melanoma skin cancers. Moreover, it is difficult to know what parts of the solar spectrum play a role in genesis of the cancer, as the mechanism has not been worked out. DNA damage is probably involved, which would imply a role for UV-B, but it might not be the only or the major contributing factor. The recently discovered role of UV-A in non-melanoma cancers gives good reason to be careful about drawing hasty conclusions.

Another problem in making risk assessments for melanoma cancers is that there are no good models in experimental animals that can be used to accurately mimic the human disease process. So far, there are only two animals that are known to get malignant melanoma from UV exposure. One is a hybrid fish, which has been created by crossing a species that gets melanoma spontaneously and another species that can get UV-induced cancer. The other animal is the American opossum.

Different research groups have used these animals to determine the action spectrum for malignant melanoma, but the two pictures that emerge are very different. In the opossum, there is no question that UV-B plays a role in the disease process, whereas UV-A can actually repair DNA damage. If humans are similar to opossums, ozone depletion would clearly increase the risk of melanoma. The fish, on the other hand, seem to be just as sensitive to UV-A as to UV-B. If this is representative of humans, there is no worry about ozone depletion causing more cancers. It does, however, cause concern over sources of UV-A as risk factors. Solariums or tanning lamps are often heavy in the UV-A part of the spectrum. Both the models also support the advice of being careful with exposure to the sun.

Another way to judge the role of UV-B is to use some limited information about the mechanisms of melanoma in people. Most of it comes from patients with the disease *xeroderma pigmentosum*. They have a deficiency in their DNA repair and are known to be at increased risk for developing melanoma. Cells from such patients show an exceptional sensitivity to UV-B radiation, whereas recent research shows that the sensitivity to UV-A is usually normal.

"These findings suggest that UV-B radiation is the critical factor for the induction of melanoma in humans. This implies that there is still reason for concern that the incidence of this dangerous type of skin cancer will increase with increasing UV-B irradiance," writes Jan van der Leun and his colleague Frank de Gruijl in a statement presented for the Parties to the Montreal Protocol in Vienna in December 1995.

A HEALTHY TAN?

Many people feel safe from the sun once they have developed a tan. It does not hurt any more because we do not get burned as easily. Maybe those of us who are naturally pale also get fooled by advertisements and other popular images of healthy tanned people. We should beware. A tan has some protective value, but it is not safe to rely on it entirely if you want to avoid skin cancer.

The darkening of the skin is caused by the pigment melanin. It is an effective absorber of UV-B and for naturally dark-skinned people it offers some real protective value. Black people very rarely get skin cancer. Rather, dark-skinned children living in sun-deprived areas can sometimes suffer from vitamin D deficiency, as the production of vitamin D in the skin requires UV-B. Without sufficient sunlight, there is a risk for the crippling disease of the skeleton called rickets.

Melanin is also involved when we get a tan, but the pigment concentrations are much lower than for naturally dark-skinned people. It is indeed so low that the tan itself is not very protective against the damaging solar rays. The protection instead comes from a parallel process in UV-exposed skin: a thickening of the epidermis, which provides an effective UV filter. UV-A does not cause this thickening of the skin and a tan from a UV-A solarium can therefore be very deceptive. There are reports of people who have gotten severe sunburns because they thought their solarium tan gave protection.

Solariums live on the notion that a tan is a sign of health. Beach vacations and coming home with a brown face from skiing evoke the same positive images. But how healthy is a tan? Recent findings present quite a different picture. The tan is actually a sign of damage. It is pieces of damaged DNA that provide the melanocytes in the skin with the signal to start producing pigment. For practical purposes, this new picture might only be a piece of a curious fact. However, it provides a sobering antidote to all the cultural messages saying that a tanned body is a healthy body.

SUNSCREENS—DECEPTIVE PROTECTION

Can sunscreens protect the skin from the cancer-causing solar rays? Sunscreens with higher and higher protection factors have come into demand as the awareness about skin cancer has grown. They can certainly keep the skin from getting sunburned, but there are several questions around how well they can prevent skin cancer. Their effect depends to a large extent on how they are used.

The major problem is the false security that sunscreens give. They protect

Figure 7.5 "Some people never learn." From a Swedish campaign to teach people to be careful in the sun. Illustration: Lars Wikfeldt, courtesy of the Swedish Cancer Foundation

well against UV-B but the body's own warning system, the sunburn, does not get activated to tell us that we have spent too much time in the sun. The risk is that the sunscreens lure us into staying in the sun for longer periods of time, which would give us the same dose of carcinogenic UV radiation as we would have received from a shorter period without a sunscreen. The dose of UV-A could even be higher.

The sunscreens with high protection factors are especially devious, according to Jan van der Leun. They are loaded with different chemicals to screen the UV radiation. The chemicals can sometimes cause allergic reactions or actually sensitize the skin to the sun, making the damage worse than without the sunscreen. Moreover, using sunscreens with high protection factors never lets the skin adapt to the sun with its normal protective mechanisms of skin thickening and tanning. If by accident an area is forgotten or the screen not applied at all, the sunburn might be worse than it would have been if the skin had adapted.

Early on there were warnings that the sunscreens in themselves could cause cancer. The reason was that some screening substances, such as PABA, are carcinogenic, but today's products generally do not contain these substances.

SOME ADVICE

In the midst of all the information about skin cancer, here and elsewhere, many people still feel confused about how to act. Some practical advice is therefore in place as a conclusion to this chapter.

- Keep away from the sun when it is most intense during the hours around noon. This is when most of the damaging UV radiation reaches the surface of the Earth.
- Remember that the sun is much stronger closer to the equator. Follow the wise local culture and take a siesta at midday.
- When swimming, be aware that the damaging UV radiation penetrates into the uppermost layer of water. You can get a sunburn even if most of your body is underwater.
- When you are on a beach, remember that sand reflects light effectively, increasing the amount of UV radiation that can reach your skin.
- Snow can almost double the skin-burning capacity of the sun, which calls for extra protection on ski trips. The radiation comes from all directions.
- Do not trust the shielding capacity of clouds. A thin cloud cover might not make that much difference.

- The shade under a tree offers some, but not complete protection, as the scattered light from the blue sky also plays a role. The amount of radiation is about half compared to the direct sun.
- Avoid burns by limiting the amount of time you stay in the sun. Sunburns have been connected to the development of the most dangerous form of skin cancer—malignant melanoma. Sunburn is not the cause of skin cancer, but it is a strong warning that you have spent too much time in the sun. And it hurts.
- Use the body's own protective mechanism by letting the skin adapt slowly to the sun. Remember that UV-A solariums will not do the job.
- Use light clothing and a wide-brimmed sun-hat to protect your skin. A white T-shirt has a protective factor of about 7. Heavier textiles, such as blue jeans, have even higher protection factors.
- Use sunscreens with care, and not as a way to extend the time you can sunbathe.
- Be especially careful to protect children against the sun. This includes them wearing clothing as well as providing places in the shade for their outdoor play.

Chapter 8

A tricky compromise

THERE IS A LONG LINE OF WOMEN WITH SMALL children on their arms in front of the health clinic. The sun is beating down from a clear sky, while they patiently wait for their turn. They come here with a hope that the vaccinations will help their children fight many of the devastating diseases in the area. How well will it work? It depends on the quality of the vaccine and the skill of the medical personnel. Now it turns out that the sun might also be a joker in the game as the ultraviolet rays have the potential to hamper rather than help the body's ability to fight infections.

The sun may also be a factor in the outcome and severity of many diseases. The painful, blistering cold sores that flare up after too much time in the sun are a well-documented example. These blisters are mostly annoying, but what are the implications if the sun can activate other viruses and parasites people harbor in the cells of their skin?

The story about the interaction between the sun and the body's immune system is still in the making, the questions far outnumber the answers, and scientific controversy abounds. But there are warning lights and, as the veil is lifted, we are starting to get a glimpse at the trade-offs of living with the sun as a constant companion. It turns out to be an evolutionary balancing act between guarding the self and recognizing foe. The mechanism has implications for skin cancer. It also raises some serious questions about tropical diseases, such as malaria and leichmaniasis. The people who will suffer the most might be those who already have trouble fighting off infections.

INTRIGUING OBSERVATIONS

The tale about UV radiation and immune suppression goes back to some observations from the early 1960s. Guinea-pigs that had been exposed to an allergen showed a lowered immunologic response after they had been irradiated with UV. The lack of reaction indicated that the immune system was doing something unusual. Another clue came from cancer research. There were reports that tumors in mice caused by ultraviolet radiation could not be transplanted to other animals. They were apparently easily recognized by the immune system of the animal getting the transplant. The obvious question was why the immune system of the irradiated mice had not fought the tumor successfully if it was so immunogenic. The question hung in the air for some time, but by the mid-1970s it was clear that the UV radiation had the same effect on the animals as X-ray treatment and chemical immunosuppressants. It compromised the immune system.

Earlier, in a completely different line of research, there had been investigations into the chemical composition of the skin. Among the compounds that had been characterized was urocanic acid (UCA). It was shown that UCA absorbs energetic photons in the UV-B wavelength range and the compound was labeled a natural sunscreen. Somewhere along the line, the news was picked up by the cosmetics industry and UCA became an ingredient in sunscreens.

The impetus to follow up on these observations came from environmental politics. Plans for supersonic transport followed by the discovery of the longevity of CFCs had put the ozone layer on the political agenda. With the depletion of ozone follows increased UV radiation and immunosuppressive effects of UV-B were suddenly highly relevant.

THE QUEST FOR AN ANTENNA

For Edward de Fabo and Frances Noonan at George Washington Medical Center, Washington DC, USA, the combination of observations from basic research and questions raised by environmental politics provided the incentive for a tedious experiment made possible by technology from the US satellite program. By using special filters, they were able to pick the UV radiation apart almost wavelength by wavelength. Their goal was to investigate what part of the light actually had the ability to change the immune system. With older technology the only choice had been to either illuminate only a very small area of a study animal or to not be very specific about the wavelengths. Now they could get a dose-response curve at 5-nanometer intervals over the entire UV-B range by studying the relative loss of immunological reaction after each irradiation treatment. Their focus was on delayed contact hypersensitivity, which is the same immune reaction people have to poison ivy and in nickel allergies. In mice, contact hypersensitivity is easily observed by measuring the swelling of the ears after applying a chemical to which the immune system normally reacts.

It took almost 6000 measurements and 2000 mice to get a picture of the action spectrum, but it turned out to be very useful. First of all they could conclude that the UV-B was absorbed in the outermost layer of the skin, the dead cell layer called the *stratum corneum*. Moreover, when de Fabo and Noonan started to match the parts of the spectrum that gave the most immunosuppression with the absorption spectra of different compounds in the skin, they found an almost perfect match—UCA, the compound previously thought of as a sunscreen.

When the radiation is absorbed by the molecule, UCA changes shape from *trans*-UCA to *cis*-UCA. This is the signal that can explain the immunosuppression, concluded de Fabo and Noonan (Figure 8.1).

Further experiments showed that the signal was not only local. If an animal was irradiated on the back, but had the ears covered, it would still have a smaller than normal immune reaction on the ears. The attempts to transplant UV-induced tumors were also followed up. The tumors were found to attach and grow on the belly of an animal that had only been irradiated on the back.

The new role of UCA was not welcome news to the cosmetics industry, which fought long and hard to keep their sunscreen formulas. The idea also kicked off a scientific controversy. Was there really enough evidence to implicate UCA in immunosuppression? Edward de Fabo and Frances Noonan thought so and have continued to investigate how the system might work.

The strongest evidence for the connection between the observed immunosuppression and UCA is the match in action spectrum as this provides a mechanism to explain how the radiation energy is converted into a biochemical signal. Other support for the idea has come from the observation that mice fed with extremely high levels of the amino acid histadine show more immunosuppression. Histadine is the chemical precursor from which the body

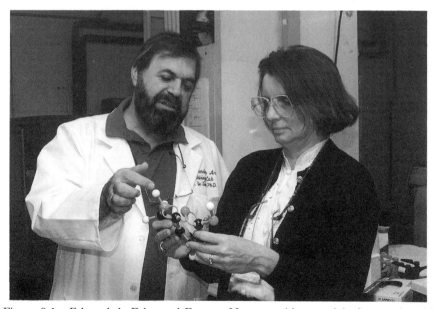

Figure 8.1   Edward de Fabo and Frances Noonan with a model of urocanic acid (UCA). Photo: Annika Nilsson

can make UCA. It also turned out that *cis*-UCA applied directly to the skin had the same effect as treatment with UV-B.

A MODEL WITH GAPS

There is no simple relationship between the conformational change of UCA and immunosuppression. In reality, a number of different cells in the skin are involved as intermediates between the signal and the response. If the proof for the involvement of this chemical antenna comes from the laboratory, the understanding of the mechanism comes from a theoretical model, which still has many gaps.

The target for the *cis*-UCA seems to be localized in the skin. It is probably a cell with which the UCA interacts. According to the model, this interaction initiates the production of something, which in turn communicates with a class of cells in the immune system called antigen-presenting cells. These have the ability to send a further message throughout the body.

The common role of the antigen-presenting cells is to take a "piece" of a foreign invader, such as a virus, process it and then show it to the rest of the immune system as a portrait of someone to watch out for and destroy. Such a presentation usually results in the production of a cell type called T-effector cells, which help the immune system to get going. The "foreign invader" could also be a skin cell that has been altered into a tumor cell, which the immune system has to destroy before it turns into a tumor.

For the immune system to work, it is just as essential that the body can recognize what is not foreign. Otherwise the risk is that it would start an immune attack on the self, an autoimmune reaction. Therefore, the antigen-presenting cells have a way of showing the immune system what not to attack by presenting a "portrait" of the nice guy, who should be left alone. In this case, the immune system starts producing T-suppressor cells that are specific for this special antigen. In the de Fabo/Noonan model of UCA-induced immunosuppression, UCA sends a signal telling the immune system that it is time to produce T-suppressor cells rather than T-effector cells.

AN EVOLUTIONARY BALANCING ACT

The down-regulation of the immune system by UV radiation probably has an evolutionary explanation. Every time light hits the skin, there is some damage to different molecules within the cells, be it DNA or proteins. If each such change caused an immune response, it would be detrimental—our immune system would constantly be attacking our skin. Since almost all animals are exposed to the sun, it would therefore be an evolutionary advantage to be able to down-regulate the immune system in response to exactly those wavelengths that cause the damage in the skin cells. However, it is a balancing act—if the system is down-regulated too much, cells that are damaged enough to turn into tumors will escape giving rise to skin cancer (Figure 8.2). One of the implications of this model is that excess exposure to UV-B caused by depletion of ozone might shift the balance to the advantage for the tumors, increasing the risk for skin cancer. In animal studies, immunosuppressive effects caused by UV-B have indeed been shown to play an important role in the outcome of both melanoma and non-melanoma skin cancers.

FINDING PIECES FOR THE PUZZLE

Some parts of the UCA model of immunosuppression have been confirmed by experiments whereas other parts are still considered black boxes. The role of the

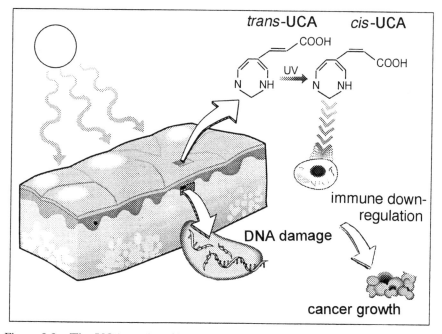

Figure 8.2   The UCA model of immune suppression. Urocanic acid (UCA) in the *stratum corneum* of the skin is converted from its *trans* to *cis* form by UV radiation. This sends a signal to the immune system to reduce its response. If the radiation has caused DNA damage, altered cells might turn into a cancer growth instead of being detected by the immune system. Illustration: Ola Rehnberg, Svenska Dagbladet

antigen-presenting cells and their sensitivity to UV radiation was fairly clear before the discovery of UCA. *In vitro* experiments have also shown that *cis*-UCA can cause an antigen that should start an immune response to be recognized as something benign that should be left alone. What is not known is how the *cis*-UCA can change the behavior of the antigen-presenting cells and that is the focus of some current research. In one experiment, radioactively-labeled UCA will be used to try to track down the receptor to which it binds. In another experiment, the hunt is on for the chemical signal that the target cell for UCA uses to communicate with the antigen-presenting cells. There is a cocktail of such immunoactive chemical signals in the body called cytokines and one or several of them might be involved.

Other experiments focus on the differences in susceptibility to skin cancer. Do different people have different levels of UCA in the skin? Studies of people with skin cancer indicate that there are individual differences linked to the risk for developing cancer. This variation might be genetic. It is well known that some skin cancers run in families. There also seems to be a sex difference for survival, with women having better chances than men. Frances Noonan is therefore looking at genetic markers on three different chromosomes in mice trying to pinpoint what the genes actually do. A third experiment deals with diet as an explanation for different levels of UCA.

CONTROVERSIAL TARGET AND A COMPETING CANDIDATE

When UCA entered the limelight in UV research in the mid-1980s, two controversies started. One, as already mentioned, was with the cosmetics industry. If UCA could hamper the immune system rather than act as a natural sunscreen, it would be the last thing you would want to smear on sun-exposed skin. But not everyone believed the compound was actually involved in immunosuppression. Along the way, the cosmetics industry did an experiment that turned out to support the hypothesis—UCA smeared on human skin does suppress contact hypersensitivity response. de Fabo says he advised against the experiment because of the ethical implications involved, but they went ahead anyway. The results, however, have never been published. Eventually, UCA was removed from sunscreens.

The controversy was also scientific in nature. The question was whether there was enough evidence to put UCA on the center stage or whether DNA should be the main actor. One of the pioneers in the work of immunosuppression, Margaret Kripke, and her coworkers have shown that immunosuppression in marsupial animals can be reversed by irradiation with UV-A. One example is the opossum. This is important as opossums have a UV-A-governed repair system for one specific type of DNA damage caused by UV-B, namely dimer formation. As UV-A can repair damage to the DNA and reverse immunosuppression, it would imply that DNA rather than UCA is the target for the energetic photons. In mammals, Kripke and her coworkers have shown that immunosuppression can be reversed in other ways that specifically repair DNA damage. Their conclusion is clear: DNA is the primary target for UV-B-induced immunosuppression.

de Fabo and Noonan counter the DNA argument by saying that this molecule might not be the only target for UV-A in the experiments and that the effects could be a collection of responses to the same radiation signal. In a current line of research they will be looking for another photoreceptor, specific for UV-A, that could block the signal from UCA. What this receptor could be no one knows. One possibility is that UV-A turns on specific genes that are responsible for the production of one or more counter signals.

By now, both UCA and DNA are usually recognized as antennas for the UV-B signal to the immune system. There might be other compounds involved as well. However, regardless of the mechanism, it might be that the balance between UV-A and UV-B determines whether the immune system recognizes a new acquaintance as friend or foe. In looking at the long-term effects of ozone depletion and in making any risk assessment for UV-induced immunosuppression, it will be essential to learn about the relative importance of the different wavelengths.

SKIN CANCER

What does UV-induced immunosuppression mean in real life? What are the health risks in an environment with increased levels of UV-B? Even if it is too early to put numbers on the risks, there is no doubt that immunosuppression is a health risk to take into consideration in our relationship to the sun. It is no doubt a contributing factor in the development of skin cancer as it increases the risk that the initial DNA damage will go undetected and a pre-cancerous cell turn into a tumor. It could be part of the explanation why tumors also develop on parts of the body that are not exposed much to the sun.

INFECTIOUS DISEASES

For other diseases than skin cancer, the signs that UV-induced immunosuppression is a significant risk factor have only started to be gathered. It probably is important, but there are only scattered animal studies to use as a base for any discussion of risk. The critical diseases would be those that the body normally fights with the cell-mediated arm of the immune system. These diseases include viral infections and parasitic diseases where the pathogen gains entrance to the body through the skin. Other parts of the immune system, such as the antibody response, are not affected by UV.

For viral infections the focus has been on herpes—a virus that among other things causes cold sores. UV-B is known to activate latent herpes simplex infections in humans. Some studies in mice show the same thing and also demonstrate that parts of the immune system are suppressed. The mice in these experiments were also less able to fight the initial infection by the virus.

In addition to the effects on the immune system, UV-B also seems to act directly on the virus, where it is harbored within the cells of the skin. These studies have only been done in cell cultures but raise further concerns about herpes and also about papilloma virus, which causes warts, and HIV. For HIV, the activation has also been shown in mice and a scary scenario is that exposure to the sun could accelerate the course of AIDS.

The picture from studies of parasitic diseases is mixed. Leishmaniasis is caused by a protozoan and is visible as lesions in sun-exposed skin. In animal studies, the skin lesions became less severe with increased levels of UV-B, but there was no decrease in the number of parasites. One hypothesis is that the UV-B might help the parasite gain entrance into the body by weakening the immune system at just the critical stage where the body would normally recognize that it was being invaded. Other parasitic diseases that might get worse are malaria and trichinosis. But one should be careful about drawing simple conslusions about all parasitic diseases. For example, immunosuppression does not seem to affect the development of schistosomiasis in an animal model. Neither does it affect the animals' ability to respond to vaccinations in the experiments that have been done.

Some bacterial and fungal infections are known to take advantage of a suppressed immune system. One is the fungus *Candida*, which is always present on the skin and on many mucous membranes, but can get out of hand when there is an imbalance in the immune system. In mice, exposure to UV-B makes the *Candida* infection more severe. Other animal studies indicated that infections with mycobacteria, which are involved in diseases such as tuberculosis and leprosy, might also get worse if the immune system is suppressed by UV radiation. For leprosy, one study of healthy people has shown that UV-B can diminish the immune response to one of the proteins that is typical of the leprosy-causing bacteria. So far, no one knows what it means for people actually suffering from leprosy, however. One of the latest findings is that mice infected with the tick-borne bacteria *Borrelia* showed less of an immune response after being irradiated with UV-B.

Will more people get sick from these diseases with a depletion of the ozone layer? Again, the picture in real life is probably complex. There are many factors that will determine the outcome of an infection. The disease-causing organism has to reach the body in the first place and the exposure depends more on climate, culture and socio-economic factors. And if a person is exposed, his or her general health might be more important than increases in UV-B for what happens next. A major problem in making any risk assessment is also that hardly anything is known about how UV radiation affects the risk for disease in humans.

VACCINATIONS

Most vaccinations depend on the immune system's ability to store information about potentially disease-causing organisms. Depending on the specific disease, different parts of the immune system are more or less important for this memory. Sometimes the body relies heavily on the antibody-based immune response, which is not at all affected by UV radiation. In other cases, the cell-mediated immune response carries a heavier burden. In such cases, there is a risk that immunosuppression might make the vaccination less effective. For example, there are some animal experiments showing that the immune response to tuberculosis vaccinations are lowered after UV exposure. If people are exposed to the sun just before they get the vaccination, one can imagine a scenario where the combination of sun and vaccine teaches the body that the mycobacterium is a nice fellow to leave alone, rather than telling it that this is something to be destroyed. The picture is obviously too simplistic to be useful in making any predictions, but it raises a warning flag. What is needed before saying anything about a potential loss of effectiveness in vaccination programs are studies of the complete immune response with the different enhancing and depressing factors that balance each other in the final outcome.

SUPPRESSION WITHOUT SUNBURN

Can the body protect itself from immunosuppression? The sunburn has always been a warning sign in our relationship to the sun, but this warning system does not seem to work when dealing with the effects on the immune system. The UV doses needed to cause immunosuppression are not very high—less than what is needed to get a sunburn—and it is the cumulative dose of the UV-B wavelengths that matters rather than an occasional high exposure. This means that being in the sun often and for extended periods increases immunosuppression, even if you do not get sunburned.

The picture gets even bleaker when looking at the potential for protection from a tan or from a naturally dark complexion. None of them seem to offer much help. Immunosuppression has been seen in both dark Australian aborigines and in fair-skinned people of Celtic descent. One explanation is that the UCA is nearer to the surface of the skin than the melanin-producing cells that give the dark color that protects cells further down. Dark-skinned people might actually be at higher risk for immunosuppression as they do not get red skin as a warning about their UV dose.

The potential for changing the odds is not very high either, if you insist on exposing your body to the sun. Sunscreens which protect well against sunburn do not stop the immunosuppression. They do, however, seem to protect against UV aggravation of herpes infections. No one really knows why sunscreens do not offer complete protection, but it is probably advisable not to rely too heavily on them. Light clothing is a good alternative. Another is to try to stay out of the sun during the hours around noon when the sun is the most intense and the levels of UV-B the highest.

Humans and other animals have evolved to deal with some of the damaging sides of the sun, but when it comes to immunosuppression it is clearly a balancing act. Too much sun, even without ozone depletion, is not good. How far ozone depletion will tip the balance to the detriment of our ability to fight disease is still an open question. However, there is no doubt about the need for caution while waiting for more knowledge about what is actually going on in the skin and in the rest of the body.

Chapter 9

Sensitive sensors of light

I T IS LATE AT NIGHT AND I CAN HARDLY MAKE OUT THE path in front of me. With my eyes wide open and my pupils dilated, I see the shapes of some trees, but no colors. Passing the same route just after sunrise, the lush green and the colorful flowers paint a peaceful scene. The veins in the leaves, the jagged edge, every detail speaks, as the rays from the low sun sift through the foliage. I look up at a bird in the top of one of the trees. As the day carries on, the sun rises higher in the sky. I can see my surroundings well and clearly, but turn my eyes down away from the sky. My face wrinkles up as I squint to protect my eyes from the intense light. My pupils are small.

The healthy eye is a sensitive sensor of light. The photons pass through the cornea and lens and accurately project an image on to the retina. My brain perceives and I see the world around me. The anatomy of my face, the eyebrows, the muscles that make me squint and those that contract the pupil, ensures that enough light, but not too much, is transmitted to the sensitive retina.

It is easy to take the ability to see our surroundings for granted but for many old people these scenes could look very different. Ageing turns the lenses of the eyes hazy. In an old eye, the hazy lens might scatter the light in all directions. Only a small portion passes through to the retina and can send a signal of an image. The scattered light confuses the picture. It starts with a blurry image but slowly deteriorates into blindness (Figure 9.1).

The clouding of the lens is called cataract and is a very common disease among elderly people. Sixteen percent of the population over 50 suffers from disabling cataract and more than half have some kind of damage to their lenses. In countries with good health care, the lens is usually replaced in routine surgery and the vision restored, but it still is a leading cause of blindness in the world. Approximately 17 million people are blind because of cataract (1985).

Will more people go blind from cataract if the ozone layer gets thinner and ultraviolet radiation increases? Will vision be impaired at a younger age?

The early risk assessments by the Environmental Effects Panel of the United Nations Environment Programme projected that each percent decrease in ozone would cause an additional 100 000–150 000 blind people in the world. Many summaries of UV radiation effects also point out cataract as a leading health risk. However, in the most recent UNEP assessment, the wording is more cautious with remarks about the role of other causes, such as poor diet and diseases, even if the rate of increased risk is reaffirmed. A more careful conclusion is also apparent in a Dutch expert report. It states that it is not scientifically justifiable to quantify the effects of UV radiation on the eye, if such effects are present under normal circumstances.

The picture seems to be getting blurrier, not only with age but with each

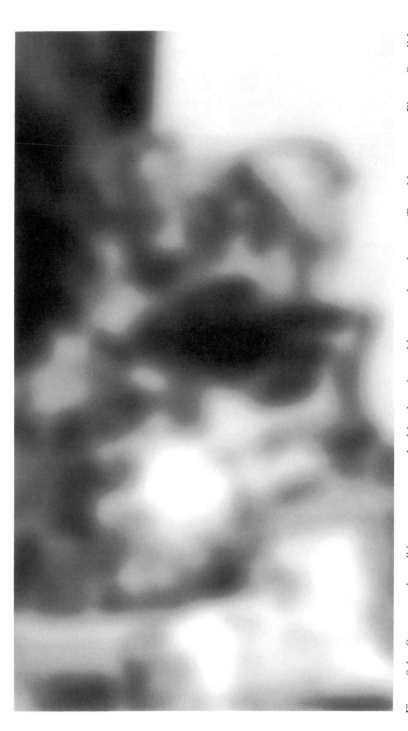

Figure 9.1 Some people walking on a street and a bicycle as it would appear through eyes affected by cataract. Photo: Swedish Association of the Visually Impaired

scientific assessment. What is going on? Maybe it is time to start looking at the basis for the risk assessment.

THE LENS

The disease of cataract affects the lens of the eye. The lens is formed as an invagination of the skin during embryonic development (Figure 9.2). The nucleus is at the center of the lens whereas the epithelium covers the front. The epithelium continues to grow throughout life, providing material for the rest of the lens. The cells elongate, lose their nucleus and other organelles and form an increasing amount of lens fiber. The major constituent of these membrane-surrounded tubes is a lens-specific protein called crystalline.

To establish a connection between UV-B and cataract, there has to be a plausible mechanism for the damage. What could cause the clouding of the lens?

Most of the evidence for light damage to the lens comes from animal experiments and work with lenses that are maintained *in vitro*. They show that light can damage the DNA as well as proteins and membranes in the cells. The damage can be caused directly by the radiation or indirectly by reactive chemicals produced by the light, such as peroxide. The result, as seen in animal studies, is that proteins change shape and aggregate into large macromolecules that can scatter light and form opaque areas in the irradiated parts. The light can also induce the production of yellowish pigments. Moreover, there is some evidence that the light can damage the growing cells in the epithelium of the lens.

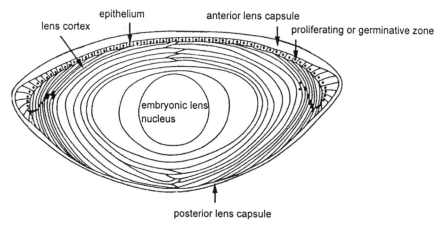

Figure 9.2   The human lens. Source: *UV Radiation from Sunlight.*

The body has several defense mechanisms against both direct and indirect damages, such as DNA repair, but it is questionable whether they protect the cells completely. That cataract develops can be taken as a sign that the repair is not sufficient. A recent finding from research on squirrel lenses also shows that UV radiation can hamper energy production, which might make the cell less able to repair other damage.

A look at the anatomy of the human eye and the current understanding of different forms of cataract indicate some possible connections between the initial damage and impaired vision.

Nuclear cataract is caused by changes in the nucleus that in most people can be detected already at the age of 30. Part of the normal ageing is that light induces the production of pigments in the cells, which causes the lens nucleus to turn yellow or even brown. Another process that starts early is that proteins in the lens nucleus react with each other forming large insoluble molecules. The result of the two processes is a decrease in the amount of light transmitted through the lens, along with a scattering of the light, which blurs the vision.

The lens cortex also experiences changes, but they normally start later in life than the yellowing of the nucleus. The changes occur mostly in the periphery of the lens, which becomes milky. In a fully developed cortical cataract, the opaque areas completely block the passage of light. The cause of these milky deposits is not fully understood. Changes in the membranes as well as in the proteins of the cells are probably involved. One difficulty in interpreting the connection to UV radiation is that the affected areas are not directly illuminated. It has raised the question as to whether the animal experiments, where the damage is in the illuminated areas, are adequate models for human cortical cataract.

The back side of the lens does not receive as much light as the front but is still subject to damage. The posterior subcapsular cataract can be described as an opaque aggregate of abnormal cells that block the light from reaching the retina. The cause might be damage to the cells in the epithelium. The damaged cells do not differentiate but migrate to the back of the lens, where they aggregate.

THE ROLE OF POPULATION-BASED STUDIES

The experimental evidence and observations of changes in the lens are not enough to show that UV-B causes cataract, but it points to some mechanism that can support such a claim. It also makes it clear that cataract is more than one disease. When looking at the further evidence, it is worthwhile to keep in mind that the changes in the lens cortex are not necessarily the same as those behind changes in the nucleus or in the back side of the lens. Previously, the diagnosis for the different types of cataract have not always been very clear, which has made it difficult to look for associations in epidemiological studies.

WHO GETS CATARACT?

The basis for quantitative risk assessments of cataract is epidemiology, where the aim is to find a connection between exposure to sunlight and the disease. A common approach is to compare the rate of disease in exposed and not-as-exposed populations. There are several such studies that indicate that people living in areas with high exposure to the sun also have a higher risk of contracting cataract. This kind of geographical study has also been used to make quantitative risk assessments.

However, simple correlations do not prove a causal link. A serious problem with all the epidemiological studies is the risk that other factors bias the results. It is well known that cataract can be connected to diabetes, malnutrition and poor hygiene. In fact, these factors are probably more important than sunlight in the overall risk assessment. Moreover, many of the geographical studies lack detailed data about each individual's dose of UV radiation.

Using a more critical approach, there are only a few studies that pass the quality criteria for good epidemiology. The most important is the Maryland Watermen study. The actors are fishermen from Chesapeake Bay, who have tried to recollect their past exposure to the sun by answering questions about work environment, leisure activities and if they wore sun-glasses or head gear. This information, along with local measurement of radiation levels at different wavelengths, has been correlated with the presence of small local opacities in the eyes of the fishermen. The result was a weak positive dose-response relationship with UV-B exposure. There was also some association with cortical cataract, but not strong enough to be statistically significant. Another study, from Beaver Dam, Wisconsin, confirmed the association between UV-B exposure and cortical cataract in men but not in women.

There is also an Italian case-control study that supports the connection between UV-B exposure and cortical cataract. People working or spending a lot of leisure time outdoors were at greater risk of getting the disease.

The connection between UV-B and posterior subcapsular cataract is not very clear. The Italian study found no correlation, but a similar investigation at Chesapeake Bay showed a connection. For nuclear cataract, several studies point to a lack of any correlation with UV-B exposure.

The epidemiological data thus suggest that of the three types of cataract, cortical cataract is the most strongly related to UV radiation. It is also the least likely to cause impaired vision.

CAREFUL CONCLUSIONS

How convincing are the experimental and epidemiological data in showing that UV-B is important for the development of cataract? Is there a basis for making quantitative risk assessment? As it turns out, there is no clear consensus among the world's experts on the disease and this is where a look at the politics behind the scene might be more informative than reading consensus statements.

First of all, there is no new scientific knowledge that could form a basis for changing the wording or risk assessment in the last compared to previous reports from the Environmental Effects Panel. The key epidemiological study was presented in 1988 and the added experimental evidence only provides a stronger connection, showing that lenses from animals that are active during the day are also damaged by UV radiation. Previous work had been done on nocturnal or twilight animals that have less need for protection.

As it turns out, the more careful wording in the recent conclusion has a quite different explanation. It is related to which people have taken part in the debate. Early on, none of the participants in the scientific review questioned the basic premise that UV-B was involved in cataract, but during recent years a different group of cataract researchers has entered the scene. This group has a completely opposite view, that UV-B is not involved at all. The critics claim that cataract in humans is caused by malnutrition and hygienic factors rather than light exposure.

In connection with the Dutch risk assessment, scientists from both views gathered at a workshop, but were still not able to reach any consensus. The chairman for the Dutch report and chairman of the UNEP Environmental Effects Panel, Professor Jan van der Leun, says that the two groups were so strongly opposed to each other's ideas that he could not let an eye expert write the workshop summary, fearing it would be biased one way or the other. The conclusions can thus be read as a balancing act by non-experts trying to make an independent judgement of the scientific evidence.

The final outcome of the workshop was the persistent lack of consensus between supporters and opponents of the sunlight hypothesis ...
On the other hand, and all participants agreed about this, it is very likely that if there is an influence of sunlight, it is probably not major compared with other factors ...

Sunlight is probably not the most important causal factor in the development of the many cataracts found in countries located in low-latitude regions.

In the literature review on cataract (as opposed to the workshop summary), the conclusion is that UV radiation probably is involved in the formation of cataract but that it is not scientifically justifiable to attempt to quantify the effects.

In another review of all of the epidemiological studies, Paul J. Dolan points out that there is only limited evidence for an association between UV-B and cortical opacities in humans and between UV radiation and posterior subcapsular cataract. Nuclear cataract is probably not caused by UV radiation at all. Dolan also cautions against campaigns to prevent cataract by protecting the eye from the sun.

The vast majority of cataract cases occur in developing countries where resources are limited. The introduction of preventive programmes aimed at reducing UV-B could divert real funds away from other effective prevention of blindness programmes.

Maybe it is time to think about the different outcomes depending on how we look at the world. When we focus our eyes on one thing, other things tend to blur into the periphery. It is only when we take a step back that we get a sense of our surroundings. And it is only when we see the whole picture that conflicting messages have a chance to merge into one. This is as true in scientific research as it is in other areas of life.

So far, UV-B risk assessments have focused only on the effects of light on the lens, but one should remember that cataract is a chronic disease which most probably has a long induction phase. What will happen if researchers look more at the eye instead of focusing on the light? What picture will develop? Maybe this wider perspective can give better insights into the role of UV-B and how it might interact with other factors. It is also worthwhile remembering that estimates of cataract risk related to UV exposure do not suggest a very strong effect, even when they are positive.

We all want the simple answers. Is ozone depletion dangerous? Yes or no? Let us hope that a more complex approach is not perceived as too difficult or demanding even though the initial picture might be a little more blurry than when we look at one thing at a time.

SNOW BLINDNESS

If cataract research is still full of questions, the picture is much more clear with acute UV damage to the eye. Many a skier might recognize the following experience:

A day on in the mountains when winter is turning to spring and the sun is high in the sky. The snow glistens as the skis break the fragile ice crystals. My glasses are safely on my nose. Evening comes, a healthy tiredness but also a feeling of grit in my eyes. It is more than Sandman trying to make me sleep. A look in the mirror shows blood-shot eyes—a reminder that the light was far stronger than it appeared. The sunglasses have saved me from a painful experience but they should have protected from all sides. The snow, not the direct sun, is the main culprit as it reflects up to 90 percent of the rays from the sun. They hit the eye from underneath, where there is no protective brow in the way.

The first target for the light when it hits the eye is the cornea, which is a half-millimeter-thick transparent tissue that covers the front of the eye. Skiing in sunny areas can cause snow blindness, which is an acute inflammation of the surface of the eyeball, or more accurately the mucous layer on the outside of the cornea. It is very painful, but the eye heals in a few days. Welders who do not use protective eye glasses are at risk of the same type of damage, which has given it its other name—welders' eye.

Would the risk of snow blindness increase with ozone depletion? Theoretically yes, as UV-B is clearly involved in the mechanism, but the major problem is in situations with large amounts of reflected light. It is easy to take precautions against snow blindness. The best advice for anyone who is out in extremely bright conditions is to use effective sunglasses that shield the light from all sides (Figure 9.3). Unlike the skin, the eye cannot adapt to strong light. It can even get more sensitive after the first damage.

LONG-TERM DAMAGE TO THE CORNEA

Ultraviolet radiation is also a contributing factor to an outgrowth of the mucous layer over the neighboring cornea called pterygium. Other parts of the cornea can degenerate as well after long exposure to UV-B. The conditions are more common in people living close to the equator and are probably aggravated by sand and salt.

WHAT HAPPENED IN PUNTA ARENAS?

Punta Arenas in southern Chile has made the headlines more than once in connection with the yearly Antarctic ozone hole. Areas of severe ozone depletion reach this southern outpost of South America and increased levels of UV radiation are a fact of life. Public education campaigns advise people to

Figure 9.3 Eye protection is important to avoid snow blindness in high UV conditions, such as the combination of snow and sunlight. Photo: Robin Round

use sunglasses and sunscreens. In the international news, there have been alarming reports of an increase in UV-related eye disease among people, and of animals going blind.

How serious are these reports? Have people in Punta Arenas actually experienced acute snow blindness because of the ozone hole? As it turns out, the situation is not very serious at all. Several teams of scientists have been to Punta Arenas to investigate the claims and one of them looked into the records of over 7000 patients who had visited their doctor during the period of ozone depletion. Of seven patients with inflammations of the cornea, five were welders who had been exposed to the bright light of welding arcs, one patient had used a UV-tanning device and one patient had an eye disorder that is generally not associated with UV-B. In their report, the team also points out that UV-B exposure in Punta Arenas, even with the increase, was less than in many other areas at the same time. None of the animals showed any sign of blinding cataract.

What is the significance of the findings in Punta Arenas? A first and obvious lesson is to take alarmist reports with a grain of salt. Looking at the UV levels should have been enough to kick suspicion into gear. If there are risks for eye disease because of ozone depletion they are probably long-term rather than acute. The exception would be people in Antarctica who do not wear good sunglasses. Unfortunately, increased rates of a certain disease 10 years in the future does not fit the journalistic style of most of the mass media. Neither does a scientific debate about how different factors, such as diet, hygiene and possibly UV-B, can contribute to disease. But stay tuned. That debate is bound to continue.

Chapter 10

Do Patagonian sheep need sunglasses?

S HEEP AND RABBITS ARE GOING BLIND IN PATAGONIA, according to headlines in *Newsweek* a few years ago. More recently, a decline in frog populations was blamed on ozone depletion. But what is the real picture? Will animals get sick if ultraviolet radiation levels increase? Which animals are sensitive?

Animal health and UV radiation have not attracted much research and the body of knowledge is far too small to make any general remarks or risk assessments. However, reports such as those from Patagonia have spurred new studies. Moreover, a new focus on ecological interactions is starting to widen the perspectives to include animals.

EYE DISEASE IN GRAZING ANIMALS

It is well-known from previous experiments and field studies that grazing cattle can develop certain eye infections when they are exposed to UV-B. There is also an association between UV-B and squamous cell carcinomas of the tissue around the eye in both cattle and horses. After the first reports of blind animals in Patagonia, a research team with veterinary ophthalmologist Kirk Gelatt went to southern Chile to investigate the reports of blinded animals together with local veterinarians. At local farms, they looked at over 200 sheep, 30 cattle and 29 alpacas.

There is no question that some animals had eye diseases. For example, most of the cattle suffered from keratoconjunctivitis, an inflammation of the cornea that is connected with chlamydia infections. However, the number of sick animals was not unexpectedly high and there was no known link between these infections and UV-B. Several animals also had signs of cataract, but none of them were blind. Kirk Gelatt concludes that the overall frequency of eye disease in the animals varied but that it was not excessive. Nor could it be related to past UV-B levels.

The study clearly shows that the first reports of animals blinded by the ozone hole were false. However, in the long run there might indeed be serious animal health effects of increased UV-B, and Kirk Gelatt has ideas about continued studies. These would involve using UV-sensitive film, which is placed under the eyes of each animal to make it possible to monitor individual UV exposure. Ten years from now, Kirk Gelatt and his collegues might be able to answer how much ozone depletion will increase the risk of eye disease in grazing animals.

SENSITIVE FROG EGGS

During the past decades the populations of some frog species have declined drastically. One cause is change in land use, which has destroyed their habitats, but some of the threatened species are in relatively undisturbed areas. Moreover, the decline is taking place all over the world. This has led to speculations about a connection to global changes, particularly ozone depletion and increases in UV radiation, as a cause behind the decline. How likely is such a connection?

To establish a cause-and-effect relationship between ozone depletion and population decline is almost impossible. There are too many different factors to take into account. However, experimental studies can point to mechanisms behind such a connection. This has been the approach in a study of frog eggs at Oregon State University.

The Oregon team had taken note that not all amphibian species were affected and hypothesized that the variation in sensitivity could be connected with differing abilities to repair DNA damage. Amphibian eggs have no hard outer shell and could be very sensitive to UV damage unless the egg had intrinsic protective mechanisms, such as efficient DNA repair. One such mechanism is the enzyme photolyase, which is activated by light and able to repair one specific kind of damage to the DNA.

After two seasons of photolyase measurements and studies of hatching success in 10 amphibian species, they saw that there were striking differences in photolyase activity. "The egg-laying behavior and photolyase activities suggest that certain amphibian species have adapted to reduce exposure of their eggs to UV-B radiation," they conclude. Salamanders, for example, hide their eggs or lay them in deep water, but there might still be enough selective pressure for efficient DNA repair to develop. Frogs that lay eggs in shallow water often had higher levels of the enzyme.

They also showed that some of the most sensitive species have the lowest photolyase activity. Among them are two that have undergone such drastic population declines that they are on their way to becoming threatened. The hatching studies show that the sun under natural conditions could be a limiting factor for their survival. They would thus be very sensitive to any increase in UV-B.

Are changes in UV-B an important cause for the decline in amphibian populations that has already taken place? This question is much more difficult to answer as it requires reliable measurements of UV-B levels. The Oregon team point at some trends over Canada with a 35 percent increase in UV radiation in the winter and 7 percent in the summer since 1989, but this is one of the few areas with documented increases in UV radiation except for Antarctica. Moreover, the figure of 35 percent has turned out not to be very representative of a general trend. What can be safely said is that there are biological mechanisms that are consistent with the UV-B hypothesis of amphibian population decline, but that this study does not exclude other causes.

FOOD FROM THE SEA

About one third of the protein people consume comes from the sea. Many developing countries are even more dependent on fishing as a source of food and threats to the productivity of the seas have been one worry with a scenario of increased ozone deletion.

The threats to aquatic animals are twofold. If phytoplankton populations decline there will be less food for animals higher up in the food chain. The second concern is that the eggs and larvae of fish and crustaceans are themselves sensitive to UV-B.

As with amphibians, the sensitivity to the sun seems to differ widely among species. For example, one type of shrimp larvae is so sensitive that half of the young die even at normal UV-B levels. Other species tolerate the radiation levels that can be expected under 16 percent ozone depletion. A closer look at mortality under different light regiments shows that the shrimp might be able to handle a certain dose, but as the DNA repair can no longer keep up there is a threshold over which damage will increase drastically.

Among vertebrates, anchovy larvae are very sensitive. After 12 days under a simulated 16 percent ozone depletion, all the larvae at half-a-meter depth along the North American Pacific coast would be dead, according to one study. Half of the larvae would be gone after only two days. Another research team has looked at rainbow trout and two threatened salmonids. The response to UV varied, but they conclude that it is an important environmental stress.

Clearly, UV radiation is something to which life in the sea has had to adapt. However, the efficiency of DNA repair and different behavioral responses sometimes lie right at the border of what is sufficient. The question is how fast they will be able to adapt further when the environment changes. If they cannot adapt, even small changes in UV-B at critical life stages might be detrimental for some species. The role for fisheries will depend on what species are affected and their role in the ecosystem.

CORAL BLEACHING

Coral bleaching has caused alarm as a sign of global warming. It is associated with warmer than normal sea water and is caused by the symbiotic algae, zooxanthella, leaving the coral tissue. Now it turns out that high levels of UV-B might be a culprit as well. In the Caribbean, the bleaching episodes in 1987 and 1990 occurred when the water was extremely calm and clear, allowing the solar rays to penetrate to greater depths than normal. Bleaching occurred as deep as 20 meters.

To investigate this possibility, a team from the University of Houston took a closer look at the corals' ability to protect themselves against extra UV radiation. Normally, the corals produce mycosporine-like amino acids, which absorb UV. The deeper the coral is located, the less of these protective compounds is present. The experiments showed that corals from deep waters were much more sensitive to extra radiation than those collected in shallow water. The team speculates that the corals' ability to adapt would not keep up with a rapid change in UV radiation, as would be caused by episodes of extremely clear and calm water. They also show that the bleaching can take place without any increase in water temperature.

Has ozone depletion played a role in the coral bleaching episodes during the past few years? The mechanism is clearly there, but the answer is probably no. The bleached corals are located in the tropics and there have been no significant changes in ozone over those areas. A risk assessment for the future would also have to take other factors into account. Changes in wind and weather patterns, which allow UV light to penetrate deeper in the water, might be more important than the state of the ozone layer.

A WEB OF INTERACTIONS

So far, most research on the effects of UV-B on animals has focused on individual species. From an ecological point of view the most far-reaching effects might instead depend on their role in the ecosystem as a whole. One example would be if differences in sensitivity would change the competitive balance between different species. Adaptation, behavior or changes in chemical defense mechanisms will also affect other creatures.

One focus on the interaction between different species within an ecosystem starts with a look at the smallest of animals: zooplankton. These tiny grazers are important links between phytoplankton and carnivores of various sizes in aquatic food webs. The sensitivity of zooplankton to UV-B has been shown in several studies, both in the laboratory and in lakes. For some species, sunlight can be lethal and they often have mechanisms to avoid the intense radiation at the water surface. When the sun is bright, it can keep them away from the warmer surface water. Many zooplankton also have protective pigments, which shows that they have had to make evolutionary investments to avoid UV damage.

The ecological implications of this sensitivity were the focus of a study that included both zooplankton and phytoplankton as well as insect larvae. A Canadian team had noticed that, against all odds, the phytoplankton increased in spite of high levels of UV-B. They could not explain the increase with different protective responses and therefore turned their attention to changes in the interaction between different organisms.

They set up the experiment at South Thompson River in Canada using the organisms from naturally flowing water. In some of the areas, light of specified wavelengths was excluded. The initial effect was that the main grazers, chironomids, avoided the areas with high levels of UV-A and photosynthetically active radiation. Later on during the experiment, the number of chironomids in UV-B-exposed areas decreased as well. It looked as if the grazers were dying off, and the response was very similar to shrimp larvae exposed to UV-B above their threshold. The most interesting observation came at the end, when there was a sharp increase in algae in the UV-B-exposed areas. The lack of grazing seemed to allow the algae to grow freely so that the biomass could continue to accumulate. The chironomids were apparently more sensitive to UV-B than the algae on which they feed. The conclusion drawn by Max Bothwell, Darren Sherbot and Cullen Pollock is that even present levels of UV-B affect the balance between primary producers and consumers in shallow bottom-dwelling communities. With increases in UV-B, changes in this balance might be more important than the effects on individual organisms.

They also point at the implications of these findings for the global effects of increased UV-B and the connection to the carbon cycle. The deleterious effects on consumers might be more important for carbon flow than the direct effect on photosynthetic algae, which has been the focus of most research so far.

It is difficult to draw any general conclusions about animal health and ozone depletion, but the message from the few studies that have been done is that UV-B is often an important environmental stress factor. As in many other areas of UV radiation research, the challenge is to understand how it interacts with other stresses. Is an increase in UV-B a major added burden for a population or an ecosystem or is it only a minor change to which life can adapt? The next chapter will address this question in an evolutionary perspective to provide a wider framework for looking at risks when we change our global environment.

Chapter 11

Evolution of knowledge

E VOLVE COMES FROM THE LATIN WORD FOR UNROLL. The dictionary of biology gives a more familiar definition: the gradual process by which the present diversity of plant and animal life arose from the earliest most primitive organisms. The biology textbook talks about evolution as all the changes that transform life on Earth.

Can evolution be used as a tool for understanding life under a thinning ozone layer? I believe it can, in two different ways: One way is by looking at ultraviolet radiation as an environmental stress to which life has had to adapt. The second way is less obvious and carries over into the philosophy of science. It is about the evolution of knowledge.

This chapter is my attempt at summarizing a journey that has taken me all the way from Abisko north of the Arctic Circle to a first-hand experience of life under the Antarctic ozone hole. Rather than conclusions based on scientific consensus, these are the pictures that have unfolded and some thoughts for yet another step on a journey. It starts with a trip back to the beginning.

A TRIP BACK TO THE BEGINNING

Time is not reckoned yet. Earth is a young planet and there is no life. It is covered by water and the atmosphere has an unfamiliar composition. No oxygen. No ozone. The radiation from the sun is fierce. Somewhere, somehow, strings of carbon and nitrogen atoms form the first nucleic acids. It is not just any molecule—it can reproduce itself. A seed of life.

This first life is vulnerable. The ultraviolet radiation can twist the molecule, lock its parts together and take away its unique ability to copy itself, its ability to carry information from one generation to the next. Still, in the deep ocean and other places away from the sun, a few survive. Enclosed by membranes, they become primitive cells. There is no need for light as the surrounding molecules supply energy for the metabolic processes.

But at the surface of the water, there is a much more abundant source of energy—the sun. Sometime between 1500 and 3000 million years ago a few cells find this unexplored habitat. Green mats of cyanobacteria soon cover the surface of some waters.

Unfortunately, the light is both a friend and foe. The bases for life—the nucleic acids—are still as sensitive to UV radiation. Exploiting the light becomes a balancing act. Any mechanism for shielding against the damaging light or any way to repair the damage would offer tremendous advantages.

The saving grace is but a by-product. Oxygen from the first photosynthesis slowly starts to fill the atmosphere. The UV rays are still fierce. They split the oxygen molecules and ozone starts to form. It is not much, counted molecule for molecule, but the change for life on Earth is fundamental. Ozone absorbs a substantial part of the UV radiation that can damage DNA. With the ozone shield, a new world opens up for life. Surface waters and land to explore with the threat of DNA damage kept in control.

The sun-basked environments are safer now than before but far from benign. Any efficient mechanism for DNA repair or compounds that screen the damaging parts of the light spectrum offer advantages. However, there is a cost. Energy has to be diverted from other functions. It becomes a matter of weighing cost against benefit—enough protection to survive and reproduce but no more.

COMMON THEMES IN PROTECTION

The diversity of life on Earth today is the result of a multitude of selective pressures offering advantages to different life forms and strategies at different times. Learning to live with the sun has obviously been one such pressure, which has acted on organisms ranging from bacteria to large mammals. Common to all seems to be the ability to repair DNA damage. It is intriguing to realize that a skin cell has to repair thousands of messed-up DNA strands for every exposure to the sun. The work done by repair enzymes in plants and plankton is probably no less. Cutting and pasting damaged DNA sequences ensures that the information is kept intact for each cell division.

Some enzymes that repair damage caused by UV-B are triggered by UV-A or visible light. This mechanism of photoreactivation should be no surprise from an evolutionary point of view. UV-B is most intense at the same time as visible light or UV-A is brightest. Photoreactivation is found in organisms ranging from blue-green bacteria to plants, fish and marsupials.

Another evolutionary strategy has been the production of UV-absorbing compounds. Flavonoids are common in plants along with anthocyanin. Fungi often contain mycosporine-like amino acids and marine organisms have other mycosporine-like compounds. Over 25 such compounds have been identified with overlapping UV-B and UV-A absorption profiles. Ultraviolet radiation of human skin induces the production of melanin.

The need to avoid excessive UV-B exposure is also obvious in structural adaptations, such as thick skin, waxy leaves or that our eyes are protected by an eyebrow. In animals, behavior to avoid intense light is one more clue that the sun can be a real foe.

ENCOURAGING AND DISCOURAGING ANSWERS

What does ozone depletion mean in this evolutionary perspective? The answers are both encouraging and discouraging. On the positive side, it shows that most life forms have some ability to adapt by short-term behavioral or chemical mechanisms. Moreover, the basis for protection is already present in the gene pool. With time, selective pressures will ensure survival. For crops and domesticated animals, the genetic diversity can be used in directed breeding programs.

The discouraging picture is that many of the systems are optimized for a light environment that has been stable for a very long time. Additional stress will cost. For some species the cost might be too high and they will go extinct or be reduced in number to the advantage of others. The variation in repair enzyme among different amphibian species gives a hint at how UV-B can threaten diversity. In Antarctica, the severe ozone depletion each spring has probably caused changes in the phytoplankton populations already.

The long-term effects on the ecosystem as a whole are difficult to predict. It might not matter at all if other less vulnerable species take over the functions of less fortunate cousins. On the other hand, if key species disappear, it could disrupt the food supply of higher animals or the decomposition of dead matter. If dimethyl sulfide (DMS)-producing plankton are more or less sensitive than other species competing for the same niche, the result might even be a change in the chemical composition of the atmosphere and the amount of clouds in the sky.

If ozone depletion is allowed to become very severe, new questions arise. The light environment will have different spectral properties compared to those that life has experienced since the blue-green bacteria started producing oxygen more than a thousand million years ago. It is a selective increase in the UV-B range and any protection triggered by other light would not be activated. We can distinguish between a bright sun at noon and the less intense light at dusk and dawn, but we cannot see if ozone is thinner at noon. Plankton are no better off than ourselves.

These discouraging answers are major reasons for applying a precautionary principle—to avoid a change in stratospheric ozone because we cannot predict the end result.

SHAKY RISK ASSESSMENTS

Much of the research on the effects of ozone depletion as well as many reviews of the research has focused on making risk assessments. Based on the mechanism for UV damage, epidemiology and productivity studies of plants and plankton, scientists have tried to put numbers on the unknown. Each percent decrease in ozone could give a 2 percent rise in skin cancer rates. A 16 percent decrease in ozone would kill almost all anchovy larvae. Other numbers give estimates of crop losses, the incidence of eye disease and primary productivity in the sea.

The previous chapters should indicate how reliable such numbers really are in each specific case. A short summary: For eye diseases, a firm statement in previous risk assessments has been moderated as new information and scientific controversy has come to the surface. For plants the trend is similar, but the background different. More careful experimental designs have produced data that make old risk assessments doubtful. The focus has shifted from productivity losses to questions about the stability of ecosystems and biodiversity. For plankton, the first numbers on effects in Antarctica were presented only a few years ago and the debate is alive about the value of extrapolating one study area to the Southern Ocean as a whole. In conclusion, there are very few effects of increased UV-B that have been satisfactorily quantified. Non-melanoma skin cancer might be the exception.

In spite of these question marks and increasing difficulties in quantifying effects, the scientific community seems to stand firm in their conclusion that the problems are severe enough to warrant a continued phase-out of ozone-depleting substances. Why? The answer is not to be found in the numbers. Again, it is the precautionary principle that has played an important role. There are enough signs that the effects can be negative even if we do not know how severe a stress every increment of ozone deletion will be.

ONE THREAT AMONG MANY

Ozone depletion is one of the few environmental problems that has received the attention of politicians world-wide and one of the few where there is international agreement about getting rid of the polluting substances. Some people call it an environmental success story. However, looking at the threats to the environment, it is important to keep a sense of perspective about the magnitude of different problems. In most cases, ozone depletion is not the worst problem today. Increased UV radiation might limit fish catches in the future, but today overfishing is a much more acute threat. The decline in frog populations probably has more to do with habitat destruction or even natural fluctuation in population size than with increases in UV. The size of crop harvests and plant productivity in natural ecosystems depend more on how well we can keep soil degradation in check. In the human realm, poor hygiene and nutrition seem to play a major role for cataract and it is our habits more than ozone depletion that determine the risk for skin cancer.

Looking at the risk assessment and the science of UV-B effects on life, I see a distressing pattern. Simple truths turn out to be not so simple. The complexity also tends to be lost in summary statements to politicians and lay people. The ifs and buts disappear when thousands of pages have to be condensed into a few sentences. As a journalist, I often find myself guilty of extracting the flashy statements that I know provoke emotions. There is a risk that the picture gets distorted.

This list of problems is not written down as an excuse for not protecting the ozone layer. Without measures against CFCs and other ozone-depleting substances, ozone depletion will continue. The extra UV radiation is likely to be an added burden, superimposed on other stresses, and thus a burden which we are much better off without. If we allow ozone depletion to become severe enough, there will indeed be a high price to pay. However, the shaky risk assessments raise some question about the evolution of knowledge and about the relationship between science and society.

POST-NORMAL SCIENCE

Science philosopher Jan Nolin recently published a dissertation in which he describes atmospheric ozone research as post-normal science. Normal science follows a pattern where the internal dynamics in each scientific field determines the current "truth," which is manifested as a dominating paradigm. Science is left to the scientists. In post-normal science, the scientific community is no longer separate from the rest of society. Political needs determine the agenda and demand answers to questions the scientists have hardly had time to pose. Applied to the science of the ozone layer, researchers find themselves trying to predict the chemical behavior of the atmosphere and to assess the risks of further emission of CFCs. It is an assessment that might never have come about unless politics had demanded it. The original discovery that CFCs could be harmful to the ozone shield could have ended up as but another scientific publication if it had not been brought to the attention of a wider audience.

Another sign of post-normal science is that different fields of science tie into each other. The atmospheric chemists can no longer work alone but strike up cooperations with those specializing in light measurements. Moreover, the external contacts become as important as the internal discussion. The message has to go out to journalists and political decision makers to incite action. Communication with scientists, other than those directly involved with the same questions, is put on a back burner.

In the light of environmental backlash, one might ask whether post-normal science has the same credibility as normal science with its internal checks and balances. Is it more likely to be driven by partisan politics than objective evaluation of the evidence? The science of ozone depletion has indeed been challenged many times in spite of the political consensus to phase-out the damaging substances.

Jan Nolin describes this backlash partly as a result of the characteristics of post-normal science. Researchers have spent more time communicating with politicians than with the scientific community and the bases for their risk assessments are not well enough known outside their own circles. However, he also emphasizes that post-normal science has its own checks and balances, which are often more efficient than those of a traditional setting. For example, the interdisciplinary approach has involved scientists with different perspectives, even if the number of people in each field is small. This has led to critical questions from more people than would normally read a scientific article. Moreover, the consensus work in formulating political recommendation has forced critical evaluations of work that would otherwise have remained unchallenged. Atmospheric chemist Sherwood Rowland has said that the political pressure has helped ensure the high quality of some experimental work that was crucial in proving the link between chlorine and ozone depletion.

Jan Nolin's analysis offers some tools to understand the dynamics of effects research as well. The early statements can be seen as risk assessments that were forced by factors external to science. Politicians needed numbers to make decisions about CFC phase-out. Unfortunately, the science of UV-B effects had not been a priority area, and in many cases there was no base to make accurate risk assessments.

The continued process opened up for a critical evaluation of the early research. The critique was often justified, especially about the type of light used in different experiments and whether it accurately mimicked ozone depletion. Discussions about action spectra for different biological processes are another step in more sophisticated risk assessment. This includes looking at the balance between damage and protective mechanisms and determining spectral weighting functions for whole organisms rather than for DNA alone. There is no question that the critical evaluation has improved the quality of work. That might not have come about as fast without the external pressures and the need to justify conclusions.

The focus on the competitive balance between organisms is another new development. Maybe this is an example of scientists having to lift their eyes from a focus on one interesting mechanism to looking at a broader system. It is also a matter of funding, as large systems studied over several years often require expensive experimental designs. One could see the pressures of post-normal science as a forced maturing process from childhood play, through adolescent grandeur to adult contemplation of complexity.

The most important lesson might be that science does not hold one truth. When the focus is on light, UV-B shows up as an important factor for the development of cataract. When the focus is different, the effects of UV-B seem smaller. The case of cataract is not only an example of how risk assessments have forced scientific controversy out into the open. It also shows that science is a process of discovery where the answers depend as much on nature as on the questions we ask. It is a constant evolution of knowledge. Some ideas hold up to debate, new tests and better experimental designs. Others do not. It is a lesson to be kept in mind when we want simple answers to complex questions.

A DIMINISHING THREAT?

Do the current answers diminish the threats associated with ozone depletion compared to the early-warning calls? In some cases the answer is yes. The increased risk for skin cancer is not as great as first predicted. Food production might not suffer as much if we can breed for UV resistance. However, the current answers also point to some serious problems, such as effects on biogeochemical cycles. Moreover, the evolutionary analysis shows that changes in biodiversity are likely, even if we do not know what they will be. Again, it becomes a matter of taking precautions against the unknown rather than evaluating known risks.

And we might never know whether the risks will be as bad as predicted. There is an international consensus about phasing-out CFCs and several other damaging substances. If the plans are carried through, the depletion of the ozone layer might not reach catastrophic proportions anywhere else than over Antarctica. The current predictions are that if everyone complies with the Montreal Protocol and its amendments the depletion will be limited to 12–13 percent over northern mid-latitudes in the winter and spring, to 6–7 percent over those latitudes in the summer and fall, and to 11 percent over southern mid-latitudes on a year-round basis. It is only 1.5–2.5 percent more than the depletion to date. With some luck, the ozone layer might be back to normal about half-way into the next century.

Will we breathe a sigh of relief if nothing visible happens and feel satisfied that we acted in time, for once? Or will we look back at the scientists of the 1980s and 1990s as the ones crying wolf when there was nothing to fear? If we do the latter, we also have to consider whether it would have been worth a global experiment to find a more definite answer. Science is never perfect. It changes over time as we pick one path rather than another: And the knowledge we gain is no more static than life itself.

Sources

From Abisko to Antarctica

Dotto, L. and Schiff, H. (1978) *The Ozone War*. Doubleday, New York.
Maduro, R. A. and Schauerhammer, R. (1992) *The Holes in the Ozone Scare*. 21st Century Science Associates, Washington, DC.

Let there be light

Effects of Increased Ultraviolet Radiation on Biological Systems (1992) Scientific Committee on Problems of the Environment, Paris.
Josefsson, W. A. P. (1993) Monitoring ultraviolet radiation. In: *Environmental UV Photobiology*, Young, A. R., Björn, L. O., Moan, J. and Nultsch, W. (eds) 73–88. Plenum Press, New York.
Madronich, S. (1993) The atmosphere and UV-B radiation at ground level. In: *Environmental UV Photobiology*, Young, A. R., Björn, L. O., Moan, J. and Nultsch, W. (eds) 1–39. Plenum Press, New York.
Madronich, S., McKenzie, R. L., Caldwell, M. M. and Björn, L. O. (1995) Changes in ultraviolet radiation reaching the earth's surface. *Ambio* **XXIV**(3): 143–152.
Scientific Assessment of Ozone Depletion: 1994 (1995) World Meteorological Organization, Global Ozone Research and Monitoring Project—Report No. 37, Geneva.
Weiler, S. and Penhale, P. A. (eds) (1994) *Ultraviolet Radiation in Antarctica: Measurements and Biological Effects*. Antarctic Research Series, Vol. 62. American Geophysical Union.

Plankton life under the ozone hole

Cullen, J. J. and Neale, P. J. (1994) Ultraviolet radiation, ozone depletion and marine photosynthesis. *Photosynthesis Research* **39**: 303–320.
Ferreyra, G. A., Schloss, I. R., Demers, S. and Neale, P. J. (1995) Phytoplankton responses to natural UV irradiance during early spring in the Weddell–Scotia confluence: an experimental approach. *Antarctic Journal of the United States* 29: (in press).
Häder, D-P., Worrest, R. C., Kumar, H. D. and Smith, R. C. (1995) Effects of increased solar ultraviolet radiation on aquatic ecosystems. *Ambio* **XXIV**(3): 174–180.
Holm Hansen, O., Lubin, D. and Helbling, E. W. (1993) UVR and its effects on organisms in the aquatic environment. In: *Environmental UV Photobiology*. Young, A. R., Björn, L. O., Moan, J. and Nultsch, W. (eds). 379–425. Plenum Press, New York.
Karentz, D. (1991) Ecological considerations of the Antarctic ozone depletion. *Antarctic Science* **3**(1): 3–11.
Neale, P. J. and Spector, A. C. (1995) UV-absorbance by diatom populations from the Weddell–Scotia confluence. *Antarctic Journal of the United States* **29**: (in press).
Smith, R. C., Prezelin, B. B., Baker, R. R., et al. (1992) Ozone depletion: Ultraviolet radiation and phytoplankton biology in Antarctic waters. *Science* **255**: 952–957.
Smith, R. C. and Cullen, J. J. (1995) Effects of UV radiation on phytoplankton. *Reviews of Geophysics Supplement* 1211–1223.

Weiler, S. and Penhale, P. (eds) (1994) *Ultraviolet Radiation in Antarctica: Measurements and Biological Effects.* Antarctic Research Series, Vol. 62. American Geophysical Union.

Sun catchers

Bornman, J. F. and Teramura, A. H. (1993) Effects of ultraviolet-B radiation on terrestrial plants. In: *Environmental UV Photobiology.* Young, A. R., Björn, L. O., Moan, J. and Nultsch, W. (eds) 427–471. Plenum Press, New York.

Caldwell, M. M., Teramura, A. H., Tevini, M., Bornman, J. F., Björn, L. O. and Kulandaivelu, G. (1995) Effects of increased solar ultraviolet radiation on terrestrial plants. *Ambio* **XXIV**(3): 166–173.

Gehrke, C., Johanson, U., Callaghan, T. V., Chadwick, D., and Robinson, C. H. (1994) The impact of enhanced ultraviolet-B radiation on litter quality and decomposition processes in *Vaccinium* leaves from the subarctic. *Oikos* 72: 213–222.

Johanson, U., Gehrke, C., Björn, L. O., Callaghan, T. V., and Sonesson, M. (1995) The effects of enhanced UV-B radiation on the subarctic heath ecosystem. *Ambio* **XXIV**(3): 106–111.

Johanson, U., Gehrke, C., Björn, L. O. and Callaghan, T. V. (1995) The effects of enhanced UV-B radiation on the growth of dwarf shrubs in a subarctic heathland. *Functional Ecology* 9: 713–719.

Elusive threads in an intricate web

Pearce, F. (1995) Iron soup feeds algas appetite for carbon dioxide. *New Scientist*, 1 July 1995: 5.

Scoy, K. V. and Coale, K. (1994) Pumping iron into the Pacific. *New Scientist*, 3 December 1994: 32–35.

Zepp, G. R., Callaghan, T. V. and Erickson D. J. (1995) Effects of increased solar ultraviolet radiation on biogeochemical cycles. *Ambio* **XXIV**(3): 181–187.

A breath of fresh air?

Tang, X. and Madronich, S. (1995) Effects of increased solar radiation on tropospheric composition and air quality. *Ambio* 24(3): 188–190.

Red alert

Boldeman, C. and Einhorn, S. (1991) *Hud i sol.* Karolinska sjukhuset LIC förlag, Stockholm.

Health Council of the Netherlands: Risks of UV Radiation Committee (1994) *UV Radiation from Sunlight.* Publication no. 1994/05E. Health Council of the Netherlands, The Hague.

van der Leun, J. C. and de Gruijl, F. R. (1993) Influence of ozone depletion on human and animal health. In: *UV-B Radiation and Ozone Depletion.* Tevini, M. (ed.) 95–123. Lewis Publishers, Boca Raton, Florida.

van der Leun, J. C. (1995) Assessment report on the NRP subtheme effects of increasing UV-B radiation. In: *Climate Change Research. Evaluation and Policy Implications.* Zwerver, S., van Rompaey, R. S. A. R., Kok, M. T. J. and Berk, M. M. (eds) 951–989. Elsevier Science.

Longstreth, J. D., de Gruijl, F. R., Kripke, M. L., Takizawa, Y. and van der Leun, J. (1995) Effects of increased ultraviolet radiation on health. *Ambio* **XXIV**(3): 153–165.

UNEP (1991) *Environmental Effects of Ozone Depletion: 1991 Update.* United Nations Environment Programme, Nairobi, Kenya.

The Dark Side of Sunlight (1993) Utrecht University, Utrecht.

A tricky compromise

Health Council of the Netherlands: Risks of UV Radiation Committee (1994) *UV Radiation from Sunlight.* Publication no. 1994/05E. Health Council of the Netherlands, The Hague.

IARC Monograph on the Evaluation of Carcinogenic Risks to Humans; Solar and Ultraviolet Radiation (1992) **55**: 175–183. International Agency for Research on Cancer, Lyon.

Jeevan, A. and Kripke, M. (1989) Effect of a single exposure to ultraviolet radiation on *Mycobacterium bovis* Bacillus Calmette-Guerin infection in mice. *The Journal of Immunology* **143**(9): 2837–2843.

Kripke, M., Cox, O., Alas, L. and Yaroch, D. (1992) Pyrimidine dimers in DNA initiate systematic immunodepression in UV-irradiated mice. *Proceedings of the National Academy of Sciences, USA* **89**: 7516–7520.

Longstreth, J. D., de Gruijl, F. R., Kripke, M. L., Takizawa, Y. and van der Leun, J. (1995) Effects of increased ultraviolet radiation on health. *Ambio* **XXIV**(3): 153–165.

Noonan, F. and de Fabo, E. (1992) Immunosuppression by ultraviolet B radiation: initiation by urocanic acid. *Immunology Today* **13**(7): 250–254.

Noonan, F. and de Fabo, E. (1993) UV-induced immunosuppression. Relationship between changes in solar UV spectra and immunological responses. In: *Environmental UV Photobiology.* Young, A. R., Björn, L. O., Moan, J. and Nultsch, W. (eds) 113–148. Plenum Press, New York.

Noonan, F. and Lewis, F. (1995) UVB-induced immune suppression and infection with *Schistosoma mansoni. Photochemistry and Photobiology.* **61**(1).

Solar UV Radiation and the Risk of Infectious Disease (1994) Summary of Workshop held on March 1–2, 1994, in Miami Beach, Florida.

Sensitive sensors of light

Dolin, P. J. (1994) Ultraviolet radiation and cataract: a review of epidemiological evidence. *British Journal of Ophthalmology* **78**: 478–482.

Environmental Effects Panel Report (1989) United Nations Environment Programme.

Health Council of the Netherlands: Risks of UV Radiation Committe (1994) *UV Radiation from Sunlight.* Publication no. 1994/05E. Health Council of the Netherlands, The Hague.

Longstreth, J. D., de Gruijl, F. R., Kripke, M. L., Takizawa, Y. and van der Leun, J. (1995) Effects of increased ultraviolet radiation on health. *Ambio* **XXIV**3): 153–165.

Schein, D. O., Vicencio, C., Muños, B., et al. (1995) Ocular and dermatolgical health effects of ultraviolet radiation exposure from the ozone hole in southern Chile. *American Journal of Public Health* **85**(4): 546–550.

Zigman, S. (1993) Ocular damage by environmental radiant energy and its prevention. In: *Environmental UV Photobiology*. Young, A. R., Björn, L. O., Moan, J. and Nultsch, W. (eds) 149–183. Plenum Press, New York.

Do Patagonian sheep need sunglasses?

Blaustein, A. R., Hoffman, P. D., Hokit, G. D., Kiesecker, J. M., Walls, S. C. and Hays, J. B. (1994) UV repair and resistance to solar UV-B in amphibian eggs: A link to population declines? *Proceedings of the National Academy of Sciences, USA* **91**: 1791–1795.

Bothwell, M. L., Sherbot, D. M. J. and Pollock, C. M. (1994) Ecosystem response to solar ultraviolet-B radiation: influence of trophic-level interactions. *Science* **265**: 97–99.

Craig, E. W., Zagarese, H. E., Schultze, P. C., Hargreaves, B. R. and Seva, J. (1994) The impact of short-term exposure to UV-B radiation on zooplankton communities in north temperate lakes. *Journal of Plankton Research* **16**(3): 205–218.

Damkaer, D. M. and Dey, D. B. (1983) UV damage and photoreactivation potentials of larval shrimp, *Pandalus platyceros*, and adult euphaisiids, *Thysanoessa raschii*. *Oecologia* **60**: 169–175.

Gleason, D. F. and Wellington, G. M. (1993) Ultraviolet radiation and coral bleaching. *Nature* **365**: 836–838.

Häder, D-P., Worrest, R. C., Kumar, H. D. and Smith, R. C. (1995) Effects of increased solar ultraviolet radiation on aquatic ecosystems. *Ambio* **XXIV**(3): 174–180.

Schein, D. O., Vicencio, C., Muños, B. et al. (1995) Ocular and dermatological health effects of ultraviolet radiation exposure from the ozone hole in southern Chile. *American Journal of Public Health* **85**(4): 546–550.

Evolution of knowledge

Nolin, J. (1995) *Vetenskapen och ozonskiktet*. Almqvist & Wiksell International, Stockholm.

Sources for illustrations

Figures have been borrowed from or adapted from the following published sources:

Ambio Vol. XXIV No. 3. Special issue on environmental effects of ozone depletion (1995) Swedish Royal Academy of Sciences. (Figures 4.4 and 5.1)

Climate Change. The IPCC Scientific Assessment. Intergovernmental Panel on Climate Change (1990) Cambridge University Press, Cambridge. (Figure 5.2)

Effects of Increased Ultraviolet Radiation on Biological Systems (1992) Scientific Committee on

Problems of the Environment, Paris. (Figures 2.2 and 7.3)

Health Council of the Netherlands: Risks of UV Radiation Committee. Publication 1994/OSE *UV Radiation from Sunlight* Health Council of the Netherlands, The Hague. (Figures 2.1, 2.9, 4.2, 7.1, 9.2)

Neale, P., et al. (1994) Ultraviolet effects on phytoplankton photosynthesis. In: *Ultraviolet Radiation in Antarctica: Measurements and Biological Effects*. Weiler, S. C. and Penhale, P. A. (eds) Antarctic Research Series Vol. 62. American Geophysical Union

Scientific Assesssmet of Ozone Depletion: 1994 (1995) World Meteorological Organization, Global Ozone Research and Monitoring Project—Report No. 37, Geneva. (Figures 1.1, 1.2, 2.3, 2.5, 2.6)

Strålning. Energi i rörelse (1987) Naturvetenskapliga forskningsrådets årsbok 1987. Swedish Natural Science Research Council, Stockholm. (Figure 2.10)

Sol dämpar immunförsvaret (1994) Svenska Dagbladet. Vetenskap. 16 Dec 1994 (Figure 8.2)

Index